Unconventional Weapons and International Terrorism

In recent years, senior policy officials have highlighted increased signs of convergence between terrorism and unconventional (CBRN) weapons. Terrorism now involves technologies available to anyone, anywhere, anytime, deployed through innovative solutions. This indicates a new and more complex global security environment with increasing risks of terrorists trying to acquire and deploy a CBRN (chemical, biological, radiological and nuclear) attack.

This book addresses the critical importance of understanding innovation and decision-making between terrorist groups and unconventional weapons, and the difficulty in pinpointing what factors may drive violence escalation. It also underscores the necessity to understand the complex interaction between terrorist group dynamics and decision-making behavior in relation to old and new technologies.

Unconventional Weapons and International Terrorism seeks to identify a set of early warnings and critical indicators for possible future terrorist efforts to acquire and utilize unconventional CBRN weapons as a means of pursuing their goals. It also discusses the challenge for intelligence analysis in handling threat convergence in the context of globalization. The book will be of great interest to students of terrorism studies, counter-terrorism, nuclear proliferation, security studies and IR in general.

Magnus Ranstorp is Research Director at the Center for Asymmetric Threat Studies at the Swedish National Defence College. **Magnus Normark** is a Research Analyst at the Division of CBRN Defence and Security at the Swedish Defence Research Agency.

Series: Political Violence
Series Editors: Paul Wilkinson and David Rapoport

This book series contains sober, thoughtful and authoritative academic accounts of terrorism and political violence. Its aim is to produce a useful taxonomy of terror and violence through comparative and historical analysis in both national and international spheres. Each book discusses origins, organizational dynamics and outcomes of particular forms and expressions of political violence.

Aviation Terrorism and Security
Edited by Paul Wilkinson and Brian M. Jenkins

Counter-Terrorist Law and Emergency Powers in the United Kingdom, 1922–2000
Laura K. Donohue

The Democratic Experience and Political Violence
Edited by David C. Rapoport and Leonard Weinberg

Inside Terrorist Organizations
Edited by David C. Rapoport

The Future of Terrorism
Edited by Max Taylor and John Horgan

The IRA, 1968–2000
An analysis of a secret army
J. Bowyer Bell

Millennial Violence
Past, present and future
Edited by Jeffrey Kaplan

Right-Wing Extremism in the Twenty-First Century
Edited by Peter H. Merkl and Leonard Weinberg

Terrorism Today
Christopher C. Harmon

The Psychology of Terrorism
John Horgan

Research on Terrorism
Trends, achievements and failures
Edited by Andrew Silke

A War of Words
Political violence and public debate in Israel
Gerald Cromer

Root Causes of Suicide Terrorism
Globalization of martyrdom
Edited by Ami Pedahzur

Terrorism versus Democracy
The liberal state response, 2nd edition
Paul Wilkinson

Countering Terrorism and WMD
Creating a global counter-terrorism network
Edited by Peter Katona, Michael Intriligator and John Sullivan

Mapping Terrorism Research
State of the art, gaps and future direction
Edited by Magnus Ranstorp

The Ideological War on Terror
World-wide strategies for counter-terrorism
Edited by Anne Aldis and Graeme P. Herd

The IRA and Armed Struggle
Rogelio Alonso

Homeland Security in the UK
Future preparedness for terrorist attack since 9/11
Edited by Paul Wilkinson et al.

Terrorism Today, 2nd Edition
Christopher C. Harmon

Understanding Terrorism and Political Violence
The life cycle of birth, growth, transformation, and demise
Dipak K. Gupta

Global Jihadism
Theory and practice
Jarret M. Brachman

Combating Terrorism in Northern Ireland
Edited by James Dingley

Leaving Terrorism Behind
Individual and collective disengagement
Edited by Tore Bjørgo and John Horgan

Unconventional Weapons and International Terrorism
Challenges and new approaches
Edited by Magnus Ranstorp and Magnus Normark

Unconventional Weapons and International Terrorism

Challenges and new approaches

Edited by Magnus Ranstorp and Magnus Normark

Routledge
Taylor & Francis Group
LONDON AND NEW YORK

First published 2009
by Routledge
2 Park Square, Milton Park, Abingdon, Oxon OX14 4RN

Simultaneously published in the USA and Canada
by Routledge
270 Madison Ave, New York, NY 10016

Routledge is an imprint of the Taylor & Francis Group, an informa business

Typeset in Times by Wearset Ltd, Boldon, Tyne and Wear
Printed and bound in Great Britain by TJI Digital, Padstow, Cornwall

British Library Cataloguing in Publication Data
A catalogue record for this book is available from the British Library

Library of Congress Cataloging in Publication Data
A catalog record for this book has been requested

ISBN10: 0-415-48439-1 (hbk)
ISBN10: 0-203-88195-8 (ebk)

ISBN13: 978-0-415-48439-8 (hbk)
ISBN13: 978-0-203-88195-8 (ebk)

Contents

Illustrations

Figures

Tables

Biographies

Editors

Magnus Normark is a Research Analyst at the Division of CBRN Defence and Security at the Swedish Defence Research Agency (FOI).

Magnus Ranstorp is Research Director at the Center for Asymmetric Threat Studies the Swedish National Defence College. He has researched extensively on Islamic extremism and terrorism over the last 20 years.

Contributors

Gary Ackerman is currently the Research Director at the National Consortium for the Study of Terrorism and Responses to Terrorism (START), a US Homeland Security Center of Excellence, based at the University of Maryland and funded by the Department of Homeland Security.

Walter Biederbick is the acting head of the Federal Information Centre on Biological Safety at the Robert Koch Institute, Germany.

Charles D. Ferguson is the Philip D. Reed Senior Fellow for Science and Technology at the Council on Foreign Relations in Washington, DC. His research focuses on prevention of nuclear and radiological terrorism, nonproliferation and analysis of nuclear energy issues.

Brian Fishman is an Assistant Professor and acting Director of Research at the Combating Terrorism Center within the US Military Academy at West Point, New York. His primary research interest is al-Qaeda in Iraq.

James J.F. Forest is Director of Terrorism Studies and Associate Professor of Political Science at the US Military Academy at West Point, where he leads educational programs for the Combating Terrorism Center. He has published ten books and dozens of articles on terrorism, homeland security and related topics, and has earned degrees from Georgetown University, De Anza College, Stanford University and Boston College.

Nancy K. Hayden is Senior Science Advisor at the Advanced Systems and Concepts Office, Defence Threat Reduction Agency in Fort Belvoir, Virginia.

Gabriele Kraatz-Wadsack is the Chief of the Weapons of Mass Destruction Branch, Office for Disarmament Affairs at the United Nations in New York, where she leads the work of supporting the activities of the United Nations in the area of weapons of mass destruction (nuclear, chemical and biological weapons), including the threat of use of weapons of mass destruction in terrorist acts, as well as missiles.

Amy E. Smithson is a Senior Fellow at the Monterey Center for Nonproliferation Studies in Washington, DC. Amy specializes in in-depth field research on issues related to chemical and biological weapons proliferation, threat-reduction mechanisms, defense, and homeland security.

Anne Stenersen is a Research Fellow at the Norwegian Defense Research Establishment's (FFI) Terrorism Research Group. Anne's research on militant Islamism is focused on CBRN (chemical, biological, radiological and nuclear) terrorism, al-Qaeda's use of the internet, and the Taliban insurgency.

Gregory F. Treverton is Director of the RAND Corporation's Center for Global Risk and Security, located in Santa Monica. His recent work has examined terrorism, intelligence and law enforcement, with a special interest in new forms of public–private partnership.

Introduction

Detecting CBRN terrorism signatures – challenges and new approaches

Magnus Ranstorp and Magnus Normark

In September 2007, German authorities announced they had foiled a major terrorist attack against American and German targets around the Frankfurt area. Three men were arrested in the so-called Operation Alberich, caught mixing chemical ingredients at a vacation house in the Sauerland region allegedly intended for use in three separate car bombs, targeting the Ramstein US military base; a nightclub; and Frankfurt International Airport.[1] What astonished investigators was not only that the three-man cell continued their attack preparations despite knowing that they were under surveillance, but also the sheer quantities of hydrogen peroxide amassed: 730 kilograms.

This incident illustrates the ready availability of vast amounts of industrial chemicals to anyone, including terrorists. One member of the terrorist cell attempted to purchase more highly concentrated hydrogen peroxide, but failed as he did not have the required permit. Instead they bought, on five occasions, 12 containers of 35 percent hydrogen peroxide which they intended to purify and distill to 75 percent using starch derived from flour.[2] The cell had also acquired 26 military-style detonators. If successful this cell would have launched terror attacks more lethal than the 2004 Madrid or 2005 London attacks combined. This incident raises the question of why this terror cell decided on using an explosive precursor and did not move in the direction of acquiring toxic chemicals.

The toxic scenario emerged with the adoption of chlorine cylinders in vehicle-borne improvised explosive devices (IEDs) by terrorists within the Iraqi insurgency, starting in October 2006. A month earlier, Abu Hamza al-Muhajir, leader of al-Qaeda in Iraq, had urged scientists to assist insurgents in the development of unconventional weapons for use against US military facilities in Iraq.[3] Over the next nine months, al-Qaeda and affiliated groups used chlorine gas attacks on at least 15 occasions against the US military, Iraqi forces and civilians. Then these types of chlorine attacks suddenly ceased. It is clear that chlorine was widely used for purifying drinking water in Iraq and as a disinfectant.[4] What is a mystery is why it suddenly disappeared given that this type of attack greatly amplifies the fear factor in the West. Will this type of tactical attack repertoire eventually migrate to Western cities in the same way that IED construction has migrated from the Iraqi to the Afghan operational theatre?

Both these incidents from Germany and Iraq illustrate the crucial importance of understanding innovation and decision-making within the context of threat convergence between terrorist groups and unconventional (CBRN) weapons and the difficulty in pinpointing what factors may drive violence escalation. It also underscores the necessity of understanding the complex interaction between terrorist group dynamics and decision-making behavior in relation to old and new technologies.

The very fact of threat convergence between terrorism and CBRN elevates the nature of the threat from a tactical level to a strategic dimension; it is quite remarkable how little corollary intellectual effort has gone into further understanding or identifying when, under what circumstances and how this convergence process will occur. The starting point for this book originated from a threefold observation.

First, senior policy officials have more and more underscored the increased signs of this convergence between terrorism and CBRN weapons without divulging the exact contours of what the threat looks like. Aum Shinrikyo; September 11 attacks by al-Qaeda; AQ Khan network and nuclear technology proliferation; anthrax; weak and failed states; new technologies available to anyone, anywhere, anytime; all these factors have been raised as indicators of a new and more complex global security environment that indicates it is not a question of if, but when, a terrorist group deploys a major CBRN attack. Scores of Western officials publicly warn us that the day is creeping closer when terrorists will use weapons of mass destruction of unconventional means. Even former UN chief Kofi Annan is somber in his outlook, as exemplified by his address to the Madrid conference announcing the UN counter-terrorism strategy: "Nuclear terrorism is still often treated as science fiction. I wish it were." Few government officials or analysts are privileged enough to have insights into the spectrum of early warning signals that collectively worry world leaders. Many analysts operate from an incomplete operational picture or from old, static assumptions devoid of precision or crucial details. Nevertheless, most public assessments are uniformly grim, as illustrated by Germany's Foreign Minister Joschka Fischer, stating that "the use of nuclear weapons by terrorists would not only result in a major humanitarian tragedy, but also would most likely move the world beyond the threshold for actually waging a nuclear war."[5] What lies behind these apocalyptic warnings, and what is the real likelihood of terrorists using CBRN weapons?

Second, most conferences dealing with the issue of threat convergence between CBRN and terrorism often end up with over-generalizations or stove-piped discussions between scientists as to what is technically possible and feasible, on the one hand, and social scientists trying to guess intentions by drawing on a limited repertoire of case studies involving CBRN incidents and terrorism, on the other. This leads not only to the methodological problem of mixing vastly different contexts and terrorist groups in an indiscriminate fashion, but also to rather imprecise guesswork. A striking feature of most conferences or workshops trying to address CBRN terrorism is how seldom they are successful in pushing the envelope of knowledge by finding fruitful avenues of convergence between the hard sciences and social and behavioral sciences. How do you create synergies between different disciplines' understanding of threat convergence from a multitude of analytical perspectives?

Third, the academic community has made relatively few notable advances in furthering our understanding of threat convergence. As identified by Gary Ackerman, a recent survey of all WMD terrorism publications indicated that the field has "reached something of an 'interpretative impasse'" that is reminiscent of the problems associated with early terrorism studies research with a small, closed, epistemological community and the recycling of the same material (usually Aum Shinrikyo) and the same assumptions about the trajectory of procurement, weaponization and deployment.[6] There are notable exceptions where valuable contributions have been made in the past, but very few studies provide groundbreaking avenues for further understanding of this complex issue. Even fewer studies exists that question how we know what we know, as several reported CBRN terrorism cases are fictitious or highly questionable in terms of the scientific facts surrounding the actual chemical or biological agents involved.

This threefold observation is illustrative of the kaleidoscope of major methodological challenges involved in understanding the threat convergence between CBRN and terrorism. This approach became the starting point for our intellectual effort of trying to push the envelope of knowledge in certain research areas, trying to bridge the necessary technical dimensions with a better granulated assessment of terrorist intentions. This proved to be by its very nature a difficult but necessary exercise.

This book is the cumulative result from an international workshop on CBRN and terrorism threat convergence, held at the Swedish National Defence College August 10–13, 2007, where we gathered relevant and recognized international expertise from academia and government. A number of central overarching questions emerged as critical for our assembled research team to bear in mind:

- To what degree are there substantial indications that CBRN weapons are becoming attractive methods for terrorists?
- What does CBRN terrorism entail in relation to incentives to escalate or enhance the effects of violent means and what are the necessary levels of competencies, funds, logistics and other resources for building a capacity and setting the preconditions for a successful attack?
- What are useful methodological approaches to capture the increasing complexity and technological change in relation to CBRN terrorism pathways?

As eloquently underscored by Nancy K. Hayden from Sandia National Laboratories, "to understand the evolutionary dynamics of motivations, the interplay between competing motivations of actors and how trajectories toward or away from WMD proliferation are influenced by state actions, there needs to be more extensive research into second order effects."[7]

What we do know is that contemporary terrorist groups have been thinking long and hard on the prospects of acquiring CBRN weapons and that leading Salafist-Jihadist leaders have made efforts to influence networks or agents of influence worldwide to pursue this capacity by extensive rhetoric and information campaigns justifying the pursuit of escalated effects of violence, indiscriminate killings and the use of CBRN weapons. This visible elevation of CBRN into the realm of the Salafist–Jihadist discourse suggests that there is at least a possibility that a range of non-state actors may pursue a path toward acquiring CBRN weapons capability for terror purposes.

No matter how likely or unlikely one may perceive the CBRN terrorism threat to be, the potential consequences of such an event in combination with the cascading effects of technological development and horizontal proliferation necessitates the development of a methodological approach to better analytically capture the multidimensional and complex nature of this threat conversion. Furthermore, development of new mechanisms to sense various capacities for groups moving in this direction is necessary, given the tremendous consequences should these groups decide to pursue a CBRN path in the future.

What is meant by the term "CBRN"?

We have chosen to use the term CBRN in our study objectives to take into account all uses of chemical, biological, radiological and nuclear substances and materials for mass-impact terrorist purposes. The term "weapons of mass destruction" (WMD) has to a large extent become a political one with a bad appellation, especially after the US arguments for the attack on Iraq in 2003. In general, the term WMD often refers to classical chemical, biological and nuclear agents and materials, weaponized through some form of militarily significant carrier system which, for most part, is in the hands of so-called "rogue states." As such, the term "WMD" does not distinguish enough between the vast spectrum of different technical aspects that exists within this category of violent means and methods. Strictly speaking, the only category of weapon capable of *true* mass destruction is nuclear weapons. These significant differences have been taken into account in this study as we deal with the individual categories of agents and materials, in regards to their accessibility, effects and inherent properties for terrorist purposes, organized and discussed separately in individual chapters.

The potential factors influencing terrorists' approach to violent means and methods, on the other hand, are slightly different in character and have to a large extent common grounds over the wide spectrum of CBRN tracks. At the center of these aspects lie the possible *incentives* and *disincentives* to change or develop the toolbox for violent means from the current, conservative and traditional modus operandi in regards to the choice of hardware for terror attacks. Closely connected to this is the general notion of enhancing the psychological effect that choosing CBRN options entails in comparison to bullets and explosives.

Structure and approach

The focus of this research study is to explore new and innovative ways forward in better understanding and predicting the potential convergence between CBRN weapons[8] and terrorism. The starting point (and inspiration) for this approach is the critical perspective of Gary Ackerman who, based on a meticulous state-of-the-art review of the literature and knowledge base, concludes that WMD terrorism research has reached "an interpretative impasse."[9] The past may not be a good indicator for future epochal leaps in technological evolution and how terrorist groups view and assess these weapon options. New approaches are necessary to consider convergence of different knowledge bases to better address the complexity of the variables involved.

This study has adopted an approach aimed to understand the multidimensional aspects of non-state actors: their group dynamics, incentives and potential capacity-building signatures in the process of acquiring and deploying a CBRN terror capacity (see Figure I.1). The main challenge in this approach is to look beyond the few and, from the terrorist groups' perspective, relatively unsuccessful CBRN terror attempts that have been made so far. The principal aim is to unlock the potential innovative capabilities of a non-state actor for creating a "successful" mass-impact CBRN-based terror incident.

From the top-down perspective (see Figure I.1), this study seeks to understand the overlapping proliferation environments as they move toward the non-state actor/group perspective. What factors and early warning signals exist within this type of technical/environment-based approach?

The principal goal of this study is to identify a set of early warnings/critical indicators for possible future terrorist efforts to acquire and utilize CBRN weapons as a means to pursue their goals. In this context some aspects can be singled out as highly relevant in order to explore details facilitating the process of

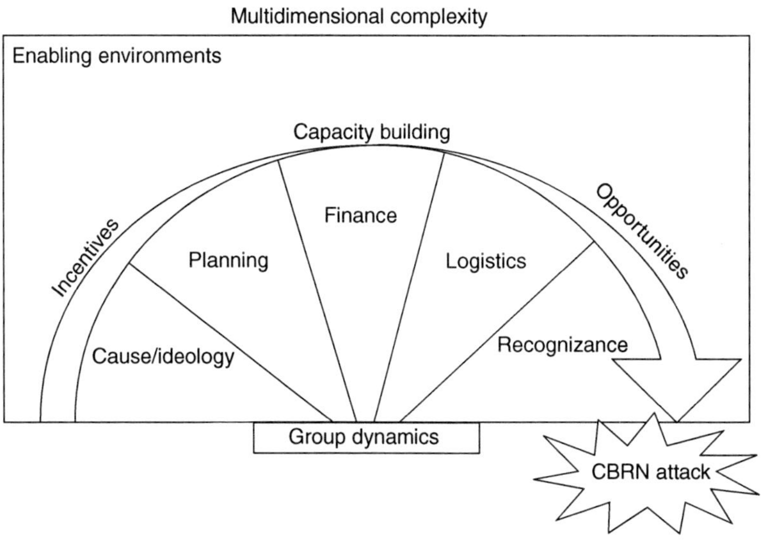

Figure I.1 The complexity of terrorist targeting versus proliferation environments.

understanding the complexity of the CBRN terrorism domain and possible future expressions of such a development.

A starting point for this study was to identify the state-of-the-art knowledge on threat convergence (CBRN terrorism) taking into account the contemporary research efforts made over the last two decades. The principal aim for such an approach was to identify possible knowledge gaps and to find new innovative approaches for future waves of research efforts.

It is clear that CBRN terrorism is not a prioritized choice for most terrorist groups in their decision to employ violence. This very fact raises avenues to explore critical factors that influence the choice of what kind of violent means are pursued to achieve their goals; what factors can push them into another direction in the future and what can impel them to broaden their toolbox to include CBRN means in their attack mode? These critical issues are at the heart of the challenge to understand the threat and, if approached in a nuanced way, will provide us with a set of potential factors to consider in monitoring pathways in this direction, in order to block or divert activities of this kind to counter such development. It is also necessary in order to build CBRN terrorism resilience in our societies, making us less vulnerable to such attacks.

Understanding terrorism behavior and motivation is also important, especially the question of what drives innovative capacity within terrorist groups in regards to violent means and modus operandi, and whether it is possible to distinguish if and to what extent ideological and political factors play any role in decisions to pursue CBRN methods.

Another significant aspect of this study is the issue of terrorists' capacity-building and the exploration of what stage in a terrorist groups' preparations for a CBRN attack is most likely to reveal critical indicators for counter-terrorism (CT) purposes. Some of the important aspects to specifically examine are the specific nature of potential indicators in terms of acquisition of technical resources and competencies and to assess the state–non-state nexus.

Which states possessing WMD capabilities (past or present programs) could potentially support non-state actors with means to carry out a CBRN mass-impact attack? Furthermore, given that there are terrorist groups around the world with an interest in acquiring CBRN capabilities for terror attacks, it is relevant to discuss what type of environment could facilitate opportunities for non-state actors to develop a CBRN weapon and prepare a CBRN attack. It's generally assumed that weak and failed states would facilitate an environment and opportunities for developing illicit CBRN means for terrorism purposes with little risk of interruption or obstruction. To what extent would the competencies, materials and equipment needed for the development of CBRN weapons be accessible in these areas? Would terrorist groups regard illicit efforts too risky to be pursued in a European country, where materials and technology are more commonly available for military, civilian or industrial use? All these issues are important to examine and could provide new valuable contributions in order to facilitate means for detection of pathways toward CBRN terrorism for the future.

After discussing the various incentives for terrorists to pursue a CBRN path toward terrorism, as well as the signatures of such efforts through indicators, this study focuses on the thorny issue of methodology to capture this complexity. How does the world of intelligence analysis approach these multifaceted and difficult threat convergence issues in their approach to uncertainty and complexity, specifically in predicting threat convergence? Are there methodological approaches which would be suitable for the intelligence community and other law enforcement functions to develop in order to understand and detect non-state actors' possible commitment to deploy a CBRN weapon for terror purposes? From the academic community and government-sponsored scientists, what are the cutting-edge questions that remain unanswered and how should one approach so-called "wicked problems" and social complexity? What should future research focus on and why?

Structure of book chapters

This book is divided into four parts and provides a rich analytical smorgasbord in asking and identifying critical questions and avenues for approaching the issue of CBRN terrorism threat convergence.

The first section of this book is dedicated to framing the field of CBRN terrorism research. Gary Ackerman of the University of Maryland provides an excellent review of the status of the current research field and outlines a series of aspects of CBRN terrorism which he argues have been given insufficient attention by the research community. He points out that the current research wave during the past five years has added little to further our insights and knowledge of threats and risks of CBRN terrorism. He argues forcefully, with striking examples of why it truly is a case for "interpretative impasse" within the field but, more importantly, navigates us through the methodological pitfalls and provides novel pathways which can be applied in order to explore new ways and methods that can advance CBRN terrorism research beyond the current status.

The second part of the book is centered on discussing the various factors influencing terrorists' attitudes toward CBRN weapons. Brian Fishman and James J.F. Forest of the Combating Terrorism Center at the US Military Academy, West Point explore terrorists' perspectives on CBRN as violent means and possible incentives for approaching the use of such weapons in the near future. The foremost focal point of this chapter is the various forms and shapes of al-Qaeda. Fishman and Forest present an insightful perception of the ideology that unites al-Qaeda and what this ideology suggests about the organization's general interest in CBRN weapons. By exploring the strategic calculations behind most terrorist groups' interest in CBRN weapons from the two factors of complexity and intended impact, they suggests an analytical framework for assessing vital factors terrorists have to consider in order to approach a CBRN pathway toward a terrorist attack. Fishman and Forest conclude their contribution by applying this analytical framework in order to assess the mode of CBRN attack most likely to fall into the respective dimensions of the al-Qaeda network.

Anne Stenersen of the Norwegian Defence Research Establishment continues the exploration of al-Qaeda-affiliated actors' viewpoint toward CBRN terrorism by critically scrutinizing the discussions and literature found on some of the radical Islamic internet sites affiliated to the al-Qaeda network. Al-Qaeda's active use of the internet over the last few years has given researchers access to some of the primary sources behind militant Islamism, including a wide range of jihadi training manuals and handbooks, of which some are dedicated to CBRN weapons. Anne's valuable research reveals important findings such as the limited consideration which the Islamists, active on these internet sites, give to the issues of CBRN as a violent means for reaching their objectives, as well as the undisputable fact that the manuals and the substance of the discussion forums found on these websites display a clear lack of basic knowledge of CBRN materials, acquisition, research and dispersion, as well as innovative capability.

In the third part of the book we move into CBRN terrorism research from a technical perspective in an effort to merge the dimensions of actors, ideologies, incentives and strategic calculations with the "hard side" of the nebulous conversion equation of the terrorism threat: CBRN means and methods.

Amy E. Smithson of the Monterey Institute's Center for Nonproliferation Studies initiates this part of the book by exploring the potential characteristics of chemical weapons (CWs) that might attract terrorists to pursue such a path for violent means. Amy highlights potential acquisition and utilization routes for terror attacks with CWs, and in that process identifies a wide range of more or less critical indicators of such activities. In this chapter Amy elaborates on potential interaction between states and terrorists in the latter's efforts to acquire CWs as well as potential technological developments and advances that may lower the current threshold for terrorists to incorporate CWs in their regular toolbox. She concludes her insightful contribution by exploring the potential shortcut for terrorists' to create chemical disasters by targeting key chemical industrial facilities, transports, etc.

Next Gabriele Kraatz-Wadsack from the United Nations WMD Branch provides an excellent discussion of normative multilateral arms control approaches, past and present, to biological non-proliferation mechanisms. She continues with a perceptive discussion on procurement of equipment and materials for biological weapons, through her experiences of lessons learned from the UN mission in Iraq. This chapter is particularly valuable for identifying critical indicators for approaches toward acquiring and utilizing biological weapons. Gabriele argues that in regards to critical indicators for bioterrorism, there are many checkpoints and few chokepoints, making such activities very hard to detect in time.

A third contribution from the technical perspective is authored by Walter Biederbick from Germany's Robert Koch Institute, who scrutinizes a selection of potential biological warfare agents for terrorist attacks. Walter critically views the accessibility of these agents and the required level of knowledge these agents demand in order to be utilized and dispersed as weapons. He concludes his contribution by elaborating on the bioterrorism threat in the future by looking at technological developments, proliferation of knowledge and means and costs for employing biological agents in terror attacks.

Charles D. Ferguson from the Council of Foreign Relations concludes the section on technical perspectives on CBRN terrorism by discussing "influence diagram analysis" as a methodological approach toward assessing the potential that terrorists will converge on the pathways of nuclear and radiological weapons. In this process Charles highlights potential influences and decisions terrorists must face and challenge on the path toward employing nuclear or radiological attacks, and through this identifies possible indicators and thresholds for such activities.

The last part of this book is dedicated to the methodological challenges approaching so-called "wicked problems," characterizing the complex dimensions of twenty-first-century national security. It examines the intelligence community's (IC) perspectives of the risk of threat convergence and the challenges of approaching the complex dilemma of detecting and assessing the CBRN terrorism threat. Greg Treverton shares his extensive experiences as a senior IC practitioner and academic at RAND, discussing the difficulties in this field of sense-making for intelligence analysts by highlighting the dramatic change in targets for the intelligence communities in regards to international terrorists and the implications and challenges it represents. Greg caps his contribution by illustrating these challenges through the prism of CBRN terrorism complexity. He outlines a Bayesian, iterative process approach that the intelligence community needs to adopt in order to adjust its methods to new complex challenges, and highlights the importance of involving a multidiscipline set of experts working under the awareness that uncertainties cannot be completely resolved.

Nancy Hayden from Sandia National Laboratories argues in her chapter that we need to include much more complex scenarios between state and non-state actors when considering motivations and interventions of terrorist activity. She discusses the merits of various analytic methodologies for handling so-called "wicked problems" such as social-science modeling and simulations, as well as state-of-the-art social-network analysis. Nancy critically reviews the merits of various methodological approaches and provides a useful menu of research questions that are urgently needed to be addressed by the scholarly and policy communities.

And finally, the editors provide a concluding chapter where they reflect on the previous chapters and discuss where we should head in our next collective and individual research efforts.

Notes

1 Mark Landler, "German Police Arrest 2 in Terrorist Plot," *New York Times*, 6 September 2007.

2 For a detailed description of the plot, see: Simone Kaiser, "How the CIA Helped Germany Foil Terror Plot," *Spiegel Online*, 10 September 2007. Available at: http://spiegel.de/international/germany/0,1518,504837,00.html.

3 Sammy Salama and Gina Cabrera-Farraj, "New Leader of Al Qaeda in Iraq Calls for Use of Unconventional Weapons Against U.S. Forces: Possible Poisoning of Iraqi Security Forces at Central Iraq Base," *WMD Insights*, November 2006. Available at: www.wmdinsights.org/I10/I10_ME1_NewLeaderAlQaeda.html.

4 Richard Weitz, Ibrahim Al-Marashi and Khalid Hilal, "Chlorine as a Terrorist Weapon in Iraq," *WMD Insights*, May 2007. Online at: www.wmdinsights.org/I15/I15_ME1_Chlorine.htm
5 Joschka Fischer, "The New Nuclear Risk," *Guardian*, 31 March 2008.
6 Gary Ackerman, "WMD Terrorism Research: Whereto from Here," *International Studies Review*, Vol. 7, No. 1 (2005): p. 137.
7 Nancy Hayden, "Terrifying Landscapes: A Study of Scientific Research into Understanding Motivations of Non-state Actors to Acquire and/or Use Weapons of Mass Destruction," Defence Threat Reduction Agency, June 22, 2007.
8 Chemical, biological, radiological or nuclear means to achieve a mass-impact terror incident in terms of physical, psychological and social effects.
9 Gary Ackerman, "WMD Terrorism Research: Whereto From Here?"

Part I

The status of CBRN terrorism research

1 Defining knowledge gaps within CBRN terrorism research

Gary Ackerman

Introduction

The past decade has borne witness to a significant rise in the profile of a security issue that lies at the confluence of the perennial global security threats of terrorism and the proliferation of unconventional weapons systems (here defined as chemical, biological, radiological and nuclear (CBRN) weapons). It is not only the general public and government officials who have been captivated by the specter of CBRN terrorism; the topic has elicited keen interest from the scholarly community, several members of which have attempted to describe and evaluate the apparent threat. In contrast to the alarmist scenarios of "weapons of mass destruction (WMD) terrorism" often conjured up by a sensationalist news media, the majority of scholarship has focused on a sober analysis of both the motivations and capabilities required for non-state actors to succeed in launching CBRN attacks.[1]

A recent survey of over 120 books, journal articles, monographs and government reports dealing with CBRN terrorism,[2] conducted by the current author and a colleague, revealed that the closest thing to a consensus amongst scholars is the following set of assertions:[3]

1 A distinction must be drawn between the use of CBRN materials to cause mass casualties (i.e. what is traditionally referred to WMD[4]) and smaller-scale uses of CBRN agents as weapons. Generally speaking, the larger the intended scale of the attack, the more difficult it is to perpetrate, and hence the lower the probability of a terrorist group successfully carrying out the attack.
2 There are significant differences associated with the difficulty of terrorist acquisition and/or use between chemical, biological, radiological and nuclear weapons, as well as their potential effects. For example, nuclear weapons are unequivocally considered as WMD, and a nuclear weapons capability is the most difficult to achieve, whereas radiological weapons are mostly weapons of social and economic disruption with limited capacity for causing casualties. The public health and economic consequences of both chemical and biological terrorism can vary dramatically.

3 There exists at least a minimal possibility that a technologically and organizationally adept terrorist organization will succeed in acquiring a CBRN weapon capable of causing mass casualties.
4 The CBRN capabilities of terrorists might be improving, both as a result of technological advances and the diffusion of knowledge.
5 There is sufficient evidence to conclude that a variety of terrorist groups and individuals espousing different backgrounds and ideologies have either considered using CBRN weapons or have attempted to acquire a CBRN weapons capability.
6 There are a wide variety of motivational incentives that could make the use of CBRN weapons attractive to terrorists, from ideological traits such as an apocalyptic worldview or technological fetishism, to operational concerns such as the ability to create mass casualties, or, most importantly, the singularly tremendous psychological impact exerted by CBRN agents. There are also a host of corresponding disincentives to using these weapons,[5] ranging from possible alienation of supporters to the lack of certainty in the scope of consequences relative to conventional weapons such as high explosives.

Yet, the radius of agreement represented by the above points is rather general and tinged with indeterminacies. Scholars disagree on many of the details that would be most useful to those tasked with protecting society and preventing acts of CBRN terrorism. For example, there is little consensus on the probability of terrorists acquiring CBRN weapons, the relative weighting of different causal factors, or the ideological and structural characteristics that distinguish would-be CBRN perpetrators.[6] These differences of opinion are at least partly the result of a dearth of usable data and partly of a lack of robust attempts to use available data to empirically validate assertions using the recognized social science methods. Furthermore, there are several important topics that have not even been broached by the academic community in a robust fashion.

Indeed, I have argued elsewhere that the scholarly and policy discussion seems to have reached something of an "interpretive impasse,"[7] with the literature increasingly beginning to recycle the same interpretations and staid shibboleths. This is not meant to denigrate several fine works in this area, but rather to acknowledge that the past five years have added little to the substantive discussion surrounding CBRN terrorism, while at the same time leaving several gaps in our knowledge largely unexplored. It is often said that the first step to addressing a problem is awareness of its existence, followed by a fuller understanding of the nature of the problem. This chapter will therefore expand upon some of the issues I have raised in previous writings by laying out what I see as some of the ground that remains to be covered, as well as describing several of the inherent difficulties associated with this endeavor. Where applicable, it will also introduce examples of studies and tools that show promise for addressing some of the existing lacunae. Far from claiming that the following constitutes a comprehensive list, the discussion merely presents personal observations of fertile areas for CBRN terrorism research.

Knowing what we don't (or can't) know

> For man does not even know his hour: like fish caught in a fatal net, like birds seized in a snare, so are men caught in the moment of disaster when it falls upon them suddenly.
>
> (Ecclesiastes 9: 12)

Whether it is admitted or not, much of the discussion regarding CBRN terrorism has a predictive pallor – one of the primary policy objectives for gaining a better understanding of CBRN terrorism (as with terrorism in general) is prescriptive, i.e. to determine the likelihood of an attack and to seek preemptive means by which to decrease this likelihood. However, in so doing we bring ourselves into the treacherous terrain of forecasting, which carries with it a number of obstacles. I will mention only two as an illustration.

The first is epistemological: we need to understand that, when dealing with certain actors and systems, there are not only those things that we do not know, but also those that we absolutely cannot know. At least since the days of Gödel,[8] philosophers and mathematicians have known that truth in some systems cannot be attained. Such concepts have only recently, however, begun to enter the social sciences and policy community, with notions of formal complexity and "wicked problems."[9] In this regard, David Snowden and Cynthia Kurtz[10] describe both the complex domain, in which patterns can emerge and be perceived but cannot be predicted, and the chaotic domain, which is devoid of cause and effect. If a threat or potential threat is situated in one of these domains, the best strategy is not to attempt to predict the specifics of an outcome, but rather to ameliorate the threat through a process of probing[11] or actions designed to restructure the environment in which the threat might arise. CBRN terrorism, with myriad interacting causes, dynamics and effects is a good prima facie candidate for being a complex or wicked problem. Before we can even begin, therefore, to expound upon the likelihood or nature of CBRN terrorism beyond the horizon of the present, we need to assess which, if any, parts of the problem are even forecastable in any practical sense. Yet, as with so many areas crucial to policy, I have seen no such "meta-analysis" of the topic of CBRN terrorism, which means that we might all be jumping blindly into an analytical black hole.

The second obstacle is the current reliance, especially in the social sciences, on the principle of extrapolating either implicitly or explicitly from past events, whether these are recorded as detailed case studies or large datasets of terrorist actions. The philosophers Thomas Hobbes and David Hume resolutely emphasized the inherent perils of induction (deriving general rules from a finite number of observations), but there are also factors specific to terrorism which complicate matters even further. Both terrorist behavior and advances in CBRN technology have shown themselves to be highly dynamic phenomena. If future developments in CBRN terrorism look very different from those of today, we must be careful not to act like the proverbial generals fighting the last war by preparing responses applicable only to the terrorists and technology of yesterday, or for that matter, today. Then there is the possibility of large, sudden and unexpected shocks to the system, what have been described variously as "Black Swans"[12] or "Wild Cards." The discovery of a cheap and simple method of synthesizing pathogens, for example, would represent a radical departure from previous trends and experience and completely alter the dynamics and probability of bioterrorism. Another complicating factor is that recorded history is an imperfect guide – we often place undue reliance on past *observables*: that is, we impute causation to those factors which we are able to measure and for which we have data. Since many less tangible aspects of past cases of terrorism are not recorded (for instance, a deceased terrorist leader's true motivation for selecting CBRN over conventional weapons as opposed to what he told his followers), empirical analysis can lead to the development of false trend models and erroneous expectations of future events. Brian Jenkins correctly emphasizes, therefore, that historical analysis provides no reliable basis for forecasting catastrophic terrorism involving CBRN.[13]

At the same time, there are many trends that are both observable and consistent, and which can serve as a guide to anticipating future threats. CBRN plots and/or uses, although departures from most past cases of "traditional" terrorism, do have antecedents in the historical record, especially since the broader strategies of terrorists remain essentially the same as those of the past. It is not in vain, therefore, to seek early-warning indicators signaling terrorist motivations and capabilities for engaging in CBRN terrorism. More work needs to be done, however, on discerning which aspects of the past record of terrorism or CBRN materials can be extrapolated and which are no longer applicable, prior to employing either quantitative or qualitative inductive techniques. For those areas where the past is unlikely to offer any useful guidance, we need to explore the use of new, non-frequentist and non-deterministic methods of analysis.[14]

Operational relevance

The second gap in CBRN terrorism research occurs at the intersection between academia and the practical process of protecting national and international security. I have argued several times[15] that research into a topic such as CBRN terrorism differs normatively from research in most other academic disciplines, in that we cannot afford to wait for the organic transition from basic research to application to policy, which often occurs at a rather leisurely pace over a period of decades. The devastating potential of contemporary terrorism and the urgency with which adequate tools to understand and counter terrorists are required mean that basic research needs to be operationalized as quickly as possible in a form that can be easily digested and used in a practical setting by analysts, investigators and policy-makers. This implies that there is a need to form direct and durable links between researchers and practitioners, however distasteful this might be to purists in the academy. What it means for researchers in practice is that they need to pay close attention to the requirements of law enforcement, intelligence and senior policy-makers as they choose which topics to research.

The difficulty arises because the real-world requirements of those tasked with protecting society against CBRN terrorism are often very specific and more fine-grained than the answers the academic world has thus far provided.

I use a topical example to demonstrate the type of answers likely to be sought by those in the operational counter-terrorism or non-proliferation arenas. There is ample evidence that al-Qaeda and its affiliates have displayed a keen and enduring interest in acquiring CBRN weapons. Between Usama bin Laden's December 1998 statement that acquiring weapons of any type, including chemical and nuclear, is an Islamic "religious duty,"[16] and Nasir al-Fahd's May 2003 *fatwa* on the permissibility of WMD, there have been over 50 reported cases of al-Qaeda attempting to acquire CBRN weapons.[17] Yet, despite the prominence of this threat, many specific questions remain not only unanswered but also unasked. To list but a few examples: In the presence of resource constraints, which of chemical, biological, radiological or nuclear weapons are jihadists likely to favor and why? Is al-Qaeda or an affiliated group more likely to engage in a small-to-medium-size CBRN attack as soon as they attain the capability or will they bide their time until they develop a genuine mass-casualty ability? What do prior jihadist behavior and attack patterns indicate about which attack modes are likely to be favored if they use CBRN weapons? I would argue that these are the types of "operational" questions that will be of greatest utility to analysts and policy-makers, and to which academic research could contribute significantly. Unfortunately, the bulk of the research thus far has either been at an operationally less helpful abstract level (e.g. "Will terrorists go nuclear?") or has not been translated into concrete guidance (e.g. in the form of early-warning indicators) for practitioners.

An example of an exploratory attempt to develop such tools is the Determinants Effecting CBRN Decisions (DECiDe) Framework,[18] which combines data on the subject under investigation (such as a designated terrorist organization) with empirically-derived inferences that elucidate the CBRN decision-making process. By taking into account such factors as an organization's history, ideology, operational objectives, life-cycle status, organizational structure, organizational dynamics, resources, operational capabilities, environmental factors and cognitive or affect-based distortions to perception and information processing, to mention only some, the DECiDe Framework[19] traces the dynamic, non-ordinal and recursive process of weapon selection decisions.[20] It does this by (1) detecting and highlighting increases in the relative attractiveness of CBRN weaponry to the terrorist group or individual and said group or individual's perceived ability to acquire a specific CBRN capability, and (2) using the terrorist group's preferences and abilities to progressively restrict the set of weaponry (e.g. CBRN) it could and would pursue.

Therefore, one of the first tasks for researchers in this area is to assist in the transition of existing findings on the subject of CBRN terrorism to operational "products" – threat assessment methodologies, early-warning indicators and behavioral profiles – that can be used by analysts less familiar with the cumulative knowledge base on CBRN terrorism.

Making the most of dubious data

It would be enough if epistemological complications were all that stood between current research and a complete understanding of CBRN terrorism. However, researchers also must contend with serious shortcomings in the data on CBRN terrorism. Since there has never been even a single unequivocal case of mass-fatality terrorism using CBRN,[21] we have to (thankfully) make do with proxy data for the dependent variable – namely plots and small-scale uses of CBRN weapons by non-state actors. However, we must be especially cautious about using such proxies, since the variance of outcomes presaged by antecedents that differ only in seemingly minor aspects can be substantial. Further, there are often problems with gathering data on the independent variables. For example, even detailed case studies are hampered by the opacity of the terrorist decision-making process; the clandestinity of terrorist groups makes close observation all but impossible and even subsequent interviews with protagonists can be marred by doubts about the subjects' veracity. Post-hoc studies further assume that the right questions are being asked by the interviewers – in many cases interviewers come from an operational perspective and are more concerned with obtaining operational information from captured terrorists (such as the whereabouts of co-conspirators or the location of the next attack) than eliciting detailed decision-making data of the type that would be useful to future researchers.

These difficulties present formidable obstacles and it is therefore not surprising that the vast majority of scholarly contributions to the CBRN terrorism problem have relied upon either hypotheses derived from the general secondary literature on terrorism or anecdotal evidence provided by a handful of prominent cases, such as the Japanese Aum Shinrikyo cult's 1995 sarin release on the Tokyo subway or the 2001 dissemination of envelopes laced with *Bacillus anthracis* spores through the US postal system. The question then becomes: Is this the best we can do with the available data? I would suggest that, despite the difficulties with the availability of CBRN terrorism data, the quantitative and qualitative methods of formal social science developed over the past century in such fields as criminology, sociology, psychology and anthropology could be applied to the data which does exist. Yet I am aware of the publication of only a single set of in-depth case studies[22] and six scholarly articles that make use of large-sample data.[23] Of these, only the two articles by Ivanova and Sandler utilize anything more than descriptive statistics. The paucity of statistical studies perhaps reflects the general absence of quantitative analysis in the terrorism literature,[24] but this is even more reason that researchers should be experimenting with every analytical tool in the scientific toolbox to wring new insights from existing data on CBRN terrorism.

In addition to quantitative tools such as regression analysis and qualitative tools such as comparative case studies, a number of other approaches used in a variety of disciplines could be adopted. These include the systematic content analysis of terrorist statements common to communications scholars and formal network analysis tools that could be employed to study CBRN logistical networks. Another useful tool would be historical–theological analyses of sacred texts and other doctrinal sources to discern ideological indicators that might promote or inhibit the use of CBRN weapons by terrorist groups. An especially promising area is the use of visual analytics, which allow for massive amounts of data in disparate datastreams to be integrated and analyzed simultaneously without experience with complex statistical packages. Two examples of visualization tools that have been applied to analyzc terrorism are Starlight, developed by the Pacific Northwest National Laboratory in the United States[25] and the Global Terrorism Database Visualization System, developed by Remco Chang and colleagues at the Southeast Regional Visualization and Analytics Center (SRVAC).[26] These software packages provide a suite of visual analytic tools that are designed to encourage the perception of hidden patterns and information "nuggets" within large amounts of heterogeneous data.

It is not only the data which presents difficulties, but also those who must use the data. While terrorism is itself a heavily interdisciplinary field, CBRN terrorism is doubly so, since it involves both technical analyses from the "hard" physical and life sciences and behavioral analyses from the "soft" social sciences and humanities. Yet it seems as if the representatives of these two communities have in some senses bifurcated research into CBRN terrorism rather than combining their talents, with the "hard" scientists largely monopolizing the vulnerability and consequence assessment, and the "soft" scientists tackling analyses of the threat itself. One must acknowledge that it is almost never an easy endeavor to bring together data and expertise even across sub-disciplines within a single parent discipline. It is likely to be even more difficult to fuse the disparate types of data and collaborate across the different research cultures and practices that prevail between the "hard" and "soft" sciences. Yet, I argue that it only through just such a marriage of the two that we will be able to deliver the synergistic analysis necessary to truly comprehend the problem.

Threat interactions

One of the most common frameworks for risk assessment is to consider a threat in terms of the value[27] and vulnerability of the asset under threat, the nature of the harm agent and the capability and motivation of the potential attacker (see Figure 1.1). In the mid-1990s, early official responses to concerns about CBRN terrorism generally tended to focus on preparing for the consequences of a CBRN attack (situated at the nexus of value, vulnerability and harm agent) and were thus weighted heavily toward worst-case scenarios. Fortunately, by the end of the 1990s scholars had drawn attention also to consideration of the motivations and capabilities of potential perpetrators.

What have received far less emphasis, however, are some of the potential interactions between the different branches of the CBRN terrorism tree. Figure 1.1 shows the obvious interactions (solid arrows) between the different elements of consequence. It is also important to realize that there exists what can be termed "perceptual" interactions (shown as broken arrows). These represent the subjective beliefs about each element of the tree on the part of the attacker. Thus, at a trivial level, an attacker's beliefs about his capabilities will impact his intention to launch an attack (and vice versa). Where this becomes more interesting is in the perceptual interactions between the consequence branch and the intention to attack. If, for instance, the attacker does not perceive that the asset holds much value for the defender (usually the target government), then he is much less likely to expend his scarce resources on attacking it, assuming that his goal is to coerce or hurt the defender. The power of this line of thinking stems from the important nuance that it is not necessarily the objective value, hazard or vulnerability that we must change to deter a CBRN attack, *but only the attacker's perception* of these elements. This is the same principle behind the home-security camera boxes with blinking lights that seem to be working but are in fact empty and do not record anything. An example, expounded by John Steinbruner,[28] amongst others, is that by properly educating our populations as to the limitations of a particular CBRN agent as a weapon (or at least its comparability to other weapon types in most circumstances) and thereby "psychologically inoculating" them, we may change the perception of our vulnerability. This may make an attack using that agent less attractive to terrorists since it denies them the terror they seek to engender. Whether any such interactions are relevant and the circumstances under which they would have deterrent value will of course require further research, but ignoring such interactions can make us blind to a host of possible measures to dissuade potential attackers from engaging in CBRN terrorism.

Dangerous dynamics and nebulous negatives

Terrorism and technology development are inherently dynamic phenomena. CBRN terrorism, which is situated at the nexus of both, can thus be expected to be anything but static. If recent trends in terrorism have taught us anything, it is that terrorists are nimble actors who can be innovative when necessary.

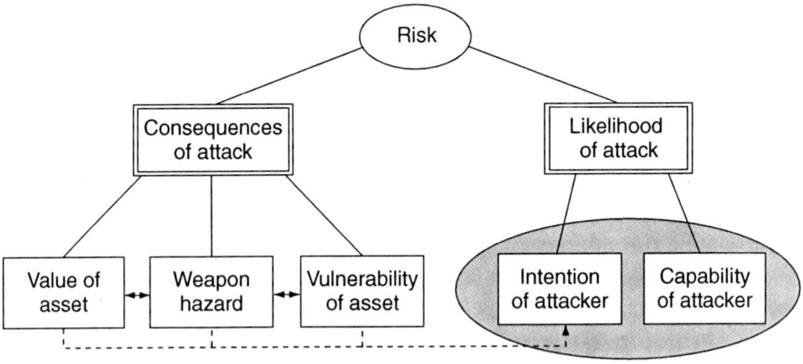

Figure 1.1 Anatomizing risk assessment of CBRN terrorism.

This is not to mention the number of new discoveries and applications being made almost daily in fields such as biology, chemistry and engineering. So, even if terrorist groups may lack the capability to engage in successful, large-scale CBRN attacks today, one would be foolish to assume that the situation will remain so forever.[29] Yet the future prospects of CBRN terrorism, both in terms of terrorist capabilities and motivations, are rarely, if ever, discussed by scholars.

Even if one believes that terrorists will eventually succeed in using WMD to cause mass casualties, the timing and rapidity of such an attack are important questions. Will this occur within the next five years or within the next 50? Still another set of issues relates to the possible demonstrative effects of a future large-scale CBRN terrorism event. If it is successful, will it encourage a string of copycats or even escalations? If its consequences are mediocre, will it have an inhibitory effect on future attacks using CBRN? Or will it merely encourage a redoubling of effort or a shift to another type of CBRN from the one that was used in the attack? I have argued in the past, and continue to do so here, for a greater emphasis on such "second order questions" in current research.

One example of promising research in this area is closer study of the possible behavior of CBRN-armed terrorists. Charles Blair has proposed research into one element of this behavior, namely Islamist terrorist command and control of nuclear weapons.[30] Blair contends that an inordinate focus amongst researchers on the requirements for current or future jihadists to gain a nuclear capability presumes that the achievement of such a capability makes employment of the weapons thus attained axiomatic. This precludes the possibility that nuclearized jihadists could conceive and implement "pragmatic and coherent nuclear weapon policies backed-up by complex and effective nuclear command and control structures."[31] Blair thus proposes research that evaluates how different jihadist groups might approach the command, control and use of nuclear weapons. For instance, would a particular strain of jihadists pre-delegate the use of nuclear weapons? What utility would they find in non-yield (i.e. non-detonation) functions for a nuclear weapon, such as deterrence, threats, etc.? Such research would draw on in-depth qualitative study of various Islamist terrorist networks, as well as the extensive prior body of knowledge on nuclear strategy, command and control gained during the decades of the Cold War.

In addition to controlling for systemic changes, any scientist worth their salt will tell you that, in order to draw defensible analytical conclusions, one should always attempt to obtain observations of variance in one's dependent variable. In our case, this would mean analyzing not only those groups or individuals who have pursued CBRN weapons, but also those who have not. It would also encourage us to focus on the question of why, with all the interest in CBRN weapons by a variety of terrorist groups, we haven't at least seen more small- or medium-size attacks. Is it a matter of insufficient capability, a satisfaction with conventional weapons or merely luck? Hypotheses abound, but very little serious research has been carried out in this area.

Conclusion

This chapter has outlined a series of aspects of CBRN terrorism to which I believe insufficient attention has thus far been paid by the research community. These range from rather abstract questions of the predictive value of CBRN terrorism research to very concrete details about the specific nature of the threat.

To sum up, more research is required:

- in conducting meta-analysis of the CBRN terrorism topic area, including which areas within the larger topic area are most amenable to predictive analysis, and which fall into the complex realm and are thus less amenable to deterministic analytical methodologies;
- for those areas least amenable to predictive analysis, to explore alternative policy measures, such as fundamentally reshaping the threat environment or performing active probes to discern the dynamics of the system;
- making use of empirical data and all the tools of the social and behavioral sciences;
- of a truly interdisciplinary nature, involving close collaboration between "hard" and "soft" scientists at all stages of the research process;
- which is designed so that its results are operationally relevant to counter-terrorism practitioners;
- that is forward-looking and explicitly takes into account the dynamics of both CBRN terrorism and the wider environment in which it might occur;
- that focuses on the majority of terrorists who do not or may not seek CBRN weapons, in addition to those that do;
- that takes into account interactions between the threat of CBRN terrorism, our vulnerability to CBRN weapons and the consequences of a CBRN terrorist attack.

I am certain that I have been given the lesser task in this volume – it is far easier to identify gaps than it is to fill them. Yet I maintain that nothing mentioned here is beyond the reach of a concerted research effort and encourage the research community to join in moving beyond the current boundaries of CBRN terrorism research.

Notes

1 See, e.g. Charles D. Ferguson, William Potter, Amy Sands, Leonard Spector and Fred Wehling, *The Four Faces of Nuclear Terrorism* (Monterey, CA: Center for Nonproliferation Studies, 2004); Arpad Palfy, "Weapon System Selection and Mass-Casualty Terrorism," *Terrorism and Political Violence* 15:2 (2003), pp. 82 and 87; Gavin Cameron, "WMD Terrorism in the United States: The Threat and Possible Countermeasures," *Nonproliferation Review* 7:1 (2000), pp. 164, 167, 169; Brad Roberts, "Has the Taboo Been Broken?" in *Terrorism with Chemical and Biological Weapons: Calibrating Risks and Responses* (Alexandria, VA: Chemical and Biological Arms Control Institute, 1997), pp. 121–40.

2 Jeffrey Bale and Gary Ackerman, *How Serious is the "WMD Terrorism" Threat? Terrorist Motivations and Capabilities for Using Chemical, Biological, Radiological, and Nuclear (CBRN) Weapons*, Report for Los Alamos National Laboratory (MIT Press, forthcoming).
3 The following list has been reproduced from Victor H. Asal, Gary A. Ackerman and R. Karl Rethemeyer, *Connections Can Be Toxic: Terrorist Organizational Factors and the Pursuit and Use of CBRN Terrorism*, unpublished manuscript (2007).
4 It is argued that the use of "weapons of mass destruction" as an umbrella term for CBRN weapons is both inaccurate (for example, not all these weapons cause "destruction," there are vast differences in the harm potential amongst these weapons and the quantification of "mass" is problematic) and counterproductive in that it may in fact increase society's sense of dread regarding CBRN and encourage our enemies to pursue these means.
5 Possibly the most systematic accounting of the various incentives and disincentives for the terrorist use of CBRN weapons can be found in Nadine Gurr and Benjamin Cole, *The New Face of Terrorism: Threats from Weapons of Mass Destruction* (New York: I.B. Tauris, 2002).
6 An explicit example of such disagreement among scholars can be found in Karl-Heinz Kamp, Joseph F. Pilat, Jessica Stern and Richard A. Falkenrath, "WMD Terrorism: An Exchange," *Survival* 40:4 (1998–1999). Others disagree on the probability of a nuclear terrorism event within the next decade, cf. Ferguson *et al.*, op. cit. and Graham Allison, *Nuclear Terrorism: The Ultimate Preventable Catastrophe* (New York: Times Books, 2004).
7 Gary Ackerman, "WMD Terrorism Research: Whereto from Here?" *International Studies Review* 7:1 (2005), p. 140.
8 Gödel's so-called Incompleteness Theorem was one of the first formal representations of this idea. See Kurt Gödel, "Über Formal Unentscheidbare Sätze der Principia Mathematica und Verwandter Systeme, I(Over Formally Undecidable Sets of the Principia Mathematica and Related Systems)," *Monatshefte für Mathematik und Physik* 38 (1931), pp. 173–98.
9 For a definition of wicked problems, see H. Rittel and M. Webber, "Dilemmas in a General Theory of Planning," *Policy Sciences* 4 (1973), pp. 155–69.
10 C.F. Kurtz and D.J Snowden, "The New Dynamics of Strategy: Sense-making in a Complex and Complicated World," *IBM Systems Journal* 42:3 (2003), pp. 462–83.
11 Akin to the echo-location of a bat, "probing" involves taking positive action within a system with the express purpose of observing the reactions of other elements of the system and thus of gaining information which is not otherwise obtainable. An example within the realm of CBRN terrorism would be to covertly "leak" or inject a distinctive recipe for creating a nerve agent into jihadist circles. Even though the recipe might not be genuine, it could be constructed to seem plausible and particularly easy to make and to provide specific signatures (such as a peculiar ingredient or process) that could be observed in the broader system. Counter-terrorism authorities could then trace the movement of the recipe through jihadist virtual and physical networks, thus increasing their information about the dissemination of this CBRN knowledge, and also identify any would-be CBRN terrorists who might try to follow the recipe.
12 Nicholas Taleb. *The Black Swan: How the Improbable Rules the World and Why We Don't Know It* (New York: Random House, 2007).
13 Brian Jenkins, "The WMD Terrorist Threat: Is There a Consensus View?" in Brad Roberts (ed.) *Hype or Reality? The "New Terrorism" and Mass Casualty Attacks* (Alexandria, VA: Chemical and Biological Arms Control Institute, 2000), pp. 242, 245.
14 For several examples of how to approach such a task, see J. Scott Armstrong (ed.), *Principles of Forecasting* (New York: Springer, 2001).

15 See, for example, Gary Ackerman, "WMD Terrorism Research: Whereto from Here?" op. cit., p. 141.
16 *Time*, December 24, 1998.
17 See "Chart of al-Qa'ida's WMD Activities," available at: http://cns.miis.edu/pubs/other/sjm_cht.htm.
18 The DECiDe Framework is being developed by the Center for Terrorism and Intelligence Studies and is based upon earlier work on assessing terrorist target selection.
19 The developers of the DECiDe Framework purposely do not refer to it as a model, since they wish to avoid the implication of a deterministic system. DECiDe merely offers a rigorous set of guidelines and will leave the ultimate conclusions in any particular case to the analysts themselves; it is a tool to facilitate and enhance, rather than replace, human analysis.
20 Recent work in the cognitive sciences suggests that decisions are often the result of numerous mental processes occurring in parallel and are more fluid in their ordering – see the theory of conceptual blending in Gilles Fauconnier and Mark Turner, *The Way We Think* (New York: Basic Books, 2002).
21 Three cases come closest to fulfilling the criteria of mass-casualty, large-consequence cases of CBRN terrorism, namely the contamination of loaves of bread with arsenic in a German POW camp in 1946 by a vengeance-seeking Jewish group named Dahm Y'Israel Nokeam; Aum Shinrikyo's 1995 Tokyo subway sarin release; and the so-called "anthrax-letter" mailings of 2001, but it is arguable whether any of these would be described as true "WMD terrorism" by most observers. More detail on the first case can be found in Ehud Sprinzak and Idith Zertal, "Avenging Israel's Blood (1946)," in Jonathan B. Tucker (ed.), *Toxic Terror: Assessing Terrorist Use of Chemical and Biological Weapons* (Cambridge, MA: MIT Press, 2000), pp. 17–41.
22 Jonathan B. Tucker (ed.), *Toxic Terror: Assessing Terrorist Use of Chemical and Biological Weapons* (Cambridge, MA: MIT Press, 2000).
23 Amy Smithson and Leslie-Ann Levy, *Ataxia: The Chemical and Biological Terrorism Threat and the US Response*. Stimson Center Report No. 35 (2000); John Parachini, "Combating Terrorism: Assessing Threats, Risk Management, and Establishing Priorities," Testimony before the House Subcommittee on National Security, Veterans Affairs, and International Relations (July 26, 2000), available at http://cns/pubs/reports/paraterr.htm, p. 2; Gavin Cameron, "WMD Terrorism in the United States: The Threat and Possible Countermeasures," *Nonproliferation Review* 7:1 (2000), pp. 163–4; Jonathan B. Tucker and Amy Sands, "An Unlikely Threat," *Bulletin of the Atomic Scientists* 55:4 (1999); Kate Ivanova and Todd Sandler, "CBRN Incidents: Political Regimes, Perpetrators, and Targets," *Terrorism and Political Violence* 18:3 (2006), pp. 423–48; Ivanova and Sandler, "CBRN Attack Perpetrators: An Empirical Study," *Foreign Policy Analysis* 3:4 (2007), pp. 273–94.
24 Cf. Leslie W. Kennedy and Cynthia M. Lum, *Developing a Foundation for Policy Relevant Terrorism Research In Criminology* (New Brunswick, NJ: Rutgers University, 2003), available from www.andromeda.rutgers.edu/~rcst/PDFFiles/ ProgressReport.doc; Alex P. Schmid, *Political Terrorism: A New Guide to Actors, Authors, Concepts, Data Bases, Theories and Literature* (New Brunswick, NJ: Transaction, 1988); Andrew Silke (ed.), *Research on Terrorism* (London: Frank Cass, 2004).
25 See http://starlight.pnl.gov for more information on the Starlight Project.
26 Xiaoyu Wang, Erin Miller, Kathleen Smarick, William Ribarsky and Remco Chang, "Investigative Visual Analysis of Global Terrorism Database," *Journal of Computer Graphics Forum* (forthcoming, 2008).
27 In the case of large-scale CBRN terrorism, determining value is fairly straightforward: the "assets" are dramatic numbers of American lives and can be regarded *ab initio* as being of high enough value to any American policy-maker to concern himself or herself with countering the threat.

28 John Steinbruner, "Terrorism: Practical Distinctions and Research Priorities," *International Studies Review* 7 (2005), p. 139.
29 This is discussed in greater detail in Ackerman, "WMD Terrorism: Whereto from Here?," op. cit., p. 142.
30 Charles Blair, *Islamist Command and Control of Nuclear Weapons*, unpublished report (2007).
31 Ibid.

Part II

Al-Qaeda motivations/ incentives for CBRN terrorism

2 WMD and the four dimensions of al-Qaeda

Brian Fishman and James J.F. Forest

Six years after 9/11, the international community faces the prospect of a catastrophic terrorist attack far larger than the attacks in New York and Washington, DC. Many experts now suggest that it is only a matter of time before al-Qaeda or some other group uses chemical, biological, radiological or nuclear (CBRN) weapons against a Western target. Countering this threat requires a clear understanding of the technical nature of catastrophic weapons and their distribution around the world, along with a comprehensive understanding of the intentions and strategy of groups and individuals that might carry out such attacks. A comprehensive counter-terrorism strategy must restrict both the capabilities of terrorists and minimize their willingness to attack.

The purpose of this chapter is to identify and explore how four primary dimensions of al-Qaeda (AQ) – AQ Central, AQ Affiliates, AQ Locals and the AQ Network – might approach the use of CBRN weapons. The discussion begins by highlighting the importance of ideology in any analysis of terrorism. We then briefly describe the four dimensions of al-Qaeda and examine the ideology that unites them, highlighting what this ideology suggests about al-Qaeda's general interest in catastrophic terrorism. Next, we explore the strategic calculation behind most terrorist groups' interests in CBRN weapons, emphasizing the two elements of complexity and intended impact, and suggest an analytical framework for assessing the opportunities and constraints a terrorist group must consider in order to optimize the costs and benefits of preparing and implementing a catastrophic terrorist attack. Finally, we use this analytical framework to assess the types of CBRN attacks most attractive to each of the four dimensions of al-Qaeda.

Understanding the contemporary threat of weapons of mass destruction (WMD) terrorism requires multiple frames of analysis, including capabilities, intended outcomes and environmental opportunities. While a great deal of attention has been focused on the latter – as evidenced by numerous global efforts to prevent WMD proliferation, along with public debate over "ungoverned spaces," "state weakness" and criminal networks in the former Soviet Union – assumptions are often made about the general utility to terrorists of a WMD attack. This chapter seeks to add granularity to those assumptions with specific regard to the four dimensions of al-Qaeda.

Ideologies of terrorism

Terrorism is an ideologically-driven phenomenon: a type of violence that transcends criminal or other motivations. Individuals and groups resort to terrorism because they have a political or religious vision of the future that they do not believe will materialize without the use of violence. Often, this vision of the future is articulated through a set of ideas and values meant to inspire individual action and rationalize the use of violence in pursuit of this envisioned future. These ideologies are often intellectually and emotionally appealing because they prescribe and explain a notion of social organization, but they also imbue individual believers with a sense of moral righteousness. This is particularly true with religious ideologies. These ideas add a spiritual dimension to the overall appeal to violence, and can thus be a more powerful motivator for extraordinary action by justifying pain and brutality via the appeal to a higher power. Setting aside the capabilities of an organization, the potential threat of a terrorist group is largely a function of how drastic the change they wish to effect is and their preferred strategy for achieving that vision.

All sorts of organizations aim to change social, political and religious reality. Most eschew violence entirely, but a small subset embraces violence for strategic or ideological reasons. An even smaller group embraces the use of massive levels of violence and unconventional weapons. It is important to recognize how ideologies that espouse mass violence represent only one small end of an ideological spectrum (see Figure 2.1). At one end of the spectrum are groups that desire dramatic changes, but do not see the necessity of violent means to bring about those changes. At various points toward the middle of this spectrum are a variety of groups willing to use some level of violence in pursuit of their objectives, ranging from a desire for religious governance (e.g. Islamic militants seeking to establish a caliphate, where sharia law reigns supreme) to Maoist communism (e.g. insurgencies in Peru and Nepal) to environmental causes (e.g. Earth Liberation Front).

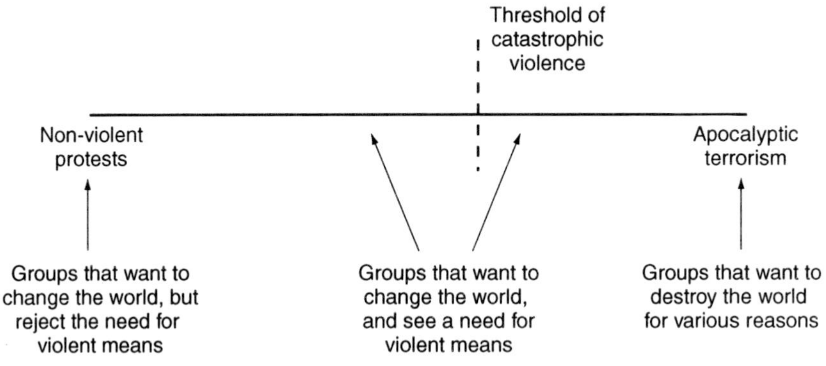

Figure 2.1 A spectrum of ideologies.

At the opposite end of the spectrum from the non-violent protestors are individuals and groups whose ideologies articulate a desire for the end of the world, or at least the end of all mankind. Examples include extreme environmentalist cults such as the Church of Euthanasia and the Voluntary Human Extinction Movement (both of whom call for the elimination of the human race in order to save the planet), and apocalyptic (doomsday or final judgment) cults. Among the latter category, the most prominent in recent years has been Aum Shinrikyo, whose leader Shoko Asahara came to believe that a catastrophic world war was imminent, and that only his followers would survive. However, it is important to note that beyond small extremist cults, there is actually a healthy tradition of worrying about the end of history within many of the world's major religions.[1] In the Bible, the Book of Revelations (also known as the Apocalypse of John) describes the eventual Battle of Armageddon between the forces of good and evil, leading to judgment day. The Qur'an does not have a Book of Revelations equivalent to provide a unified narrative about the end of the world, but there are many Muslims (particularly among the Shi'i tradition) who openly yearn and prepare for the return of the Mahdi, the messianic figure who will bring justice to the world and complete the spread of Islam. Regardless of the underlying monotheistic faith, groups that adhere to an apocalyptic ideology see their mission in two general ways: They either want to accelerate the end of time or take action to ensure that they survive the millennium. For example, Aum Shinrikyo wanted to hasten the end of the world (and thus sought nuclear weapons and developed their own chemical and biological weapons programs in pursuit of this objective),[2] while other groups have built compounds (like that of the Branch Davidians near Waco, Texas) in order to survive the apocalypse. Overall, violent groups that embrace this unique "end of times" form of catastrophic ideology represent a worst-case scenario type of terrorist threat, although to date there have been relatively few groups at this end of the spectrum (and apocalyptic groups have historically had very limited appeal to a broader population).

At a certain point along the spectrum of ideologies there is a threshold separating groups willing to engage in catastrophic terrorism from those that are not. Relatively few groups have chosen to cross that threshold. Indeed, a significant majority of terrorist groups have recognized the need to impose constraints on their violence, not only for moral reasons, but in order to maintain the popular support necessary for financing their operations and recruiting new members to their ranks. Others calibrate their violence in order to do harm and bring political change but avoid provoking an overwhelming counterattack by superior forces.

Further, many terrorists pursue a vision of the future in which they will someday be in charge of a particular governable space, and this vision may require them to overthrow an existing government but ensure that the space and people they seek to govern are left relatively undamaged. However, if the envisioned governable space is geographically separate from the hostile population of a nation-state (like a separate Basque country, a Tamil homeland, a Chechen or Kurdish state, etc.), there are fewer constraints against a catastrophic terror attack against the governing regime and those who support it (e.g. in Madrid, Moscow or Istanbul).

Among the groups that have already crossed the threshold of catastrophic terrorism, many see themselves as fulfilling the mandate of a higher power that justifies any level of violence – in essence, they reject any moral constraints on their violence. A common thread among these groups is the need for mass destruction and death (indeed, the elimination of all humans in some cases) in order to bring about a better world envisioned and articulated through some form of catastrophic ideology. However, there are also important differences. Some terrorists who adhere to catastrophic ideologies only seek reward in the afterlife, while others also want their followers to dominate this world. Members of al-Qaeda are included in this latter category.

Al-Qaeda and catastrophic terrorism

Renowned terrorism scholar Bruce Hoffman recently provided a useful way of refining our understanding of "al-Qaeda" by describing at least four dimensions of the group: al-Qaeda Central, al-Qaeda Affiliates, al-Qaeda Locals and the al-Qaeda Network.[3] "Al-Qaeda Central" refers to the organization that attacked the United States on September 11, 2001. Led by Usama bin Laden and Ayman al-Zawahiri, this organization has global reach, uses the rugged Afghanistan–Pakistan border region as a safe-haven and has demonstrated the ability to organize and implement catastrophic terrorist attacks. "Al-Qaeda Affiliates" refers to a variety of established terrorist organizations, from Jemmah Islamiyah in Indonesia to al-Qaeda in Iraq. All of these organizations have collaborated with AQ Central in some sense, though the level of cooperation varies widely. Although these organizations often espouse global goals, most have focused their militancy within the borders of a single state or region. AQ Affiliates vary widely in financial and technical resources, but al-Qaeda in Iraq, at least, can compete with AQ Central's financial resources.

"Al-Qaeda Locals" refers to small cells of individuals that have some connection to AQ Central, but no true organizational history or command and control relationship with the high command. Hoffman describes two sub-categories of AQ Locals: individuals that have some experience in terrorist organizations or training camps, and individuals that developed their own cell, but reached out to AQ Central before making their cell operational. Finally, "al-Qaeda Network" refers to people that have adopted al-Qaeda's ideology and decided independently to act in order to support al-Qaeda's overall goals. These individuals are often poorly trained and poorly equipped; there are numerous examples of AQ Networks deriving operational concepts and technical expertise from open-source materials found online.

As we demonstrate later in this chapter, Hoffman's analytical framework of four al-Qaeda dimensions is particularly useful for clarifying the true nature of the potential CBRN threat posed by followers of Usama bin Laden. But first, it is important to recognize that while each of these four dimensions operate in very different strategic environments, they are united by the same basic ideological approach to their violence, and thus understanding this ideology is critical for accurately assessing the WMD threat posed by al-Qaeda.

The ideology which unites, inspires and motivates the various forms of al-Qaeda stems from an extremist interpretation of Sunni Islam called Salafi-Jihadism. Within the 1.2 billion-strong Muslim community – people who follow the Qur'an and the example of Muhammad – there are Sunnis (people who follow the example of the Prophet) and Shi'is (people who follow the example of the Prophet and his descendents through his son-in-law Ali).[4] Salafism is a version of Sunni Islam that aims to practice Islam as its adherents believe it was practiced during and just after the time of the Prophet Muhammad. Finally, some Salafis embrace violence as the preferred method of uniting Muslims around the world and imposing a version of Islam that reflects their understanding of Islam.

This unique Salafi-Jihadi interpretation of Islam draws on a number of sources.[5] First, Ibn Taymiya – a thirteenth-century theologian – argued that Muslim leaders of his time had strayed from a literal interpretation of the Qur'an, and called for the eradication of beliefs and customs that were foreign to Islam and a renewed adherence to tawhid (oneness of God). He also embraced the concept of takfir – the expulsion of Muslims from the community of believers in order to make them legitimate targets of violence.

Ibn Taymiya's writings had great influence on Ibn Abd al-Wahhab, an eighteenth-century cleric (and founder of the religious tradition that dominates Saudi Arabia), who argued that if one could not convert an audience to his interpretation of Islam, they could be labeled as infidels and deserved to be killed. In the early twentieth century, an Egyptian named Hassan al-Banna (the founder of the Muslim Brotherhood) carried forward the argument that much of the world had fallen away from true Islam and encouraged Muslims to use violence against the corrupting influences of the West (including occupying military forces and apostate regimes). During this same period, Abu al-Ala Mawdudi – an Indian journalist – gained broad support for his argument that Muslims should not be afraid to use force in their quest to establish a more just society. Finally, another Egyptian named Sayyid Qutb expanded the argument that Islam is the one and only way of ruling mankind that is acceptable to God, and called for the abandonment of all human-created concepts, laws, customs and traditions – even by force where necessary. He argued that Muslims should resist the influences of Western institutions and traditions that have poisoned mankind and made the world an evil place (*Dar al-Harb* – house of war or chaos).

These and other prominent figures have contributed significantly to what has become known as the Salafi-Jihadist movement. The jihadis' vision of the future demands overthrowing "apostate" regimes in the Middle East and replacing them with governments that rule by Sharia law. In the pursuit of that goal, AQ Central made the strategic decision to attack Western regimes that support "apostate" governments in the Middle East. According to al-Qaeda, Western support is the "center of gravity" of their primary enemies.

Despite the widespread perception that adherents to al-Qaeda's ideology are simply bloodthirsty, the group's supporters have thought long and hard about the strategic use of weapons of mass destruction and the efficacy of catastrophic terrorism in pursuit of their political and religious goals. Jihadi articles advocating the use of such weapons tend to focus on the moral acceptability of using such weapons, whereas articles protesting catastrophic terrorism argue that they are strategically unwise. It is worth briefly surveying the internal jihadi debate over the use of catastrophic terrorism and CBRN because it illustrates both the moral and strategic calculations Salafi-Jihadi terrorists of all stripes must consider.

AQ Central and its followers have demonstrated a willingness to kill thousands of people in a single blow, which indicates they have overcome moral objections to the use of mass-casualty attacks, at least on a Western target. The ideological justifications vary widely – from the Prophet Muhammad's use of catapults to besiege cities to an exhortation from hadith to fight infidels with whatever means available. More pragmatic arguments suggest al-Qaeda should reply with the same weapons the United States used on Hiroshima and Nagasaki.

In 2003 al-Qaeda received some modicum of religious sanction for the use of WMD against the enemies of Islam by Saudi cleric Nassir bin Hamad al-Fahd, who issued an important and detailed fatwa on the permissibility of WMD in jihad.[6] Although Shaykh Nassir al-Fahd was not the first al-Qaeda supporter to advocate the use of WMD, his imprimatur was important because of his stature within the movement.

Shaykh Nassir al-Fahd stated that since America has destroyed countless lands and killed millions of Muslims, Muslims are permitted to respond in kind against the United States. He was adamant that jihadi use of WMD represents a form of symmetry with the West. As he argued that hijacking planes represents little more than a jihadi air force, Shaykh Nassir al-Fahd claimed that the US use of nuclear weapons against Japan justified using such weapons against the United States. Perhaps most importantly, Shaykh Nassir al-Fahd pointed to a saying of the Prophet Mohammad from hadith urging all Muslims to "perfectly" perform whatever actions they take, including killing. Al-Fahd interpreted the hadith to mean that modern CBRN are the most "perfect" means of killing enemies and thus are sanctioned by the Prophet.

Other followers of al-Qaeda's ideology have voiced similar kinds of arguments. AQ Central spokesman Suleiman Abu Gheith stated in 2002 that

> we have the right to kill 4 million Americans, 2 million of them children … and cripple them in the hundreds of thousands. Furthermore, it is our obligation to fight them with chemical and biological weapons, to afflict them with the fatal woes that have afflicted Muslims because of their chemical and biological weapons.[7]

A simple willingness to use WMD does not constitute a coherent strategy for employing them. Some jihadi have suggested that the timing of a WMD attack must be chosen carefully so that it achieves the political effect of controlling and intimidating the United States. In 2002 Abu Muhammad al-Ablaj noted that a chemical, biological, or nuclear weapon "must be used at a time that makes the crusader enemy beg on his knee that he does not want more strikes."[8]

In October 2006 an audio statement was released by Abu Hamza al-Muhajir – the leader of al-Qaeda in Iraq – calling for nuclear scientists to join his group, emphasizing that "the battlefield will accommodate your scientific aspirations."[9] That same month Dhiren Barot, a British jihadist and convert to Islam, pleaded guilty in a London courtroom to a series of terrorist plots in the UK and United States that were, according to authorities, "designed to kill as many innocent people as possible."[10] In one plot, Barot intended to detonate a radiological dispersion device (also known as a "dirty bomb") in London. These and other recent examples are pointed to as evidence of the global Salafi-Jihadi movement's growing interest in unconventional weapons. Contemporary members of this movement rationalize the need for these weapons as part of a power/capability/force multiplier calculation within the context of the larger socio-political vision being pursued.

Our understanding of al-Qaeda's ideology leads us to conclude that their pursuit of CBRN weaponry is designed to achieve distinct political outcomes, not herald Armageddon. Followers of this ideology – and particularly members of AQ Central, who are most concerned with attacking the far enemy – believe that WMD will advance their strategic objective of exhausting the United States economically and militarily by forcing the United States to expend massive amounts of money on protecting its critical infrastructure, borders and ports of entry, and on military deployments in Iraq, Afghanistan and elsewhere. Furthermore, AQ Central are convinced that acquiring WMD will allow al-Qaeda leaders to achieve military and strategic parity with the West, bestowing credibility on the mujahidin that might encourage more recruits to join the movement.

In an open letter to the State Department released in December 2004, senior al-Qaeda strategist Abu Musab Al-Suri argued that using WMD against the United States was the only means to fight it from a point of equality. He even criticized Usama bin Laden for not using WMD on 9/11: "If I were consulted in the case of that operation I would advise the use of planes in flights from outside the U.S. that would carry WMD."[11]

Despite the repeated threats from AQ leaders and commanders, there are reasons to believe that a catastrophic attack using CBRN weapons is not imminent. Perhaps most importantly, acquiring and delivering WMD remains a difficult task even for an organization with global reach. Critically, some al-Qaeda leaders have protested the strategy of directing large-scale attacks against the West because the provocation inevitably results in setbacks for the movement. In June 2002, a senior al-Qaeda operative argued that the 9/11 attacks had provoked a disastrous backlash that had not been worth the success of the attacks in New York and Washington, DC.[12] Even the most disastrous WMD attack on the United States would not prevent the United States from responding aggressively to such an attack. Indeed, the 9/11 attacks demonstrated the inherent weakness of al-Qaeda as much as its strength – even the tremendously "successful" attack barely disrupted the US economy and its ability to project power. Although a massive nuclear or biological attack would be much worse, little short of a widespread smallpox outbreak would fundamentally hinder the United States' ability to project power and drive world events.

In the final calculation, even WMD do not confer al-Qaeda or any other terrorist group military or strategic parity with a Great Power. In fact, many CBRN weapons – especially chemical and radiological weapons – only moderately increase the destructive power of a traditional kinetic attack. A terrorist group may obtain a tremendously destructive new weapon, but it still must make the fundamental strategic calculation of all practical political organizations: Are the benefits of this operation worth the costs? Will it alienate my supporters? Will it inadvertently empower my enemy?

For al-Qaeda, there is no single answer to those questions. The numerous formulations of militancy within the global al-Qaeda movement and the varied circumstances in which individual cells operate means that they all have unique strategic calculations. Al-Qaeda's ideology suggests that its followers are less compelled to carefully calibrate their level of violence than more traditional terrorist organizations with discrete localized demands. Despite this broad strategic latitude, al-Qaeda-linked groups must balance their violent ambitions against the capabilities of their cell, the feasibility of various attack options, and group-specific strategic considerations. In essence, we must first understand the strategic calculation of terrorists' interest in WMD before gaining quality insights into how each of the four dimensions of al-Qaeda might consider using such weapons.

The strategic calculation of WMD

Before executing any kind of attack, all terror groups must first weigh the optimal level of *complexity* of their attack and the kind of *impact* they hope to achieve. Increased complexity offers cells more operational options and, importantly, can demonstrate to a target society and observers the sophistication of the planners. Nonetheless, a complex attack is not necessarily more destructive than a simple one. Destroying a train car carrying chlorine gas in a populated area would be a relatively simple effort to release chemicals that could kill tens of thousands. Similarly, acquiring and deploying a radiological device would likely require a very complex operation to procure radiological material and develop an optimal distribution device, but the physical impact of the attack might not exceed that of a simple explosive. Just as importantly, complex attacks are much riskier than simple operations. Long logistical chains, large organizations and complex tactics increase the possibility that something in an attack cycle will go wrong.

The impact of an attack is a function of its physical and psychological effects on the target society and on observers, most importantly supporters and potential supporters of the attackers.

Al-Qaeda-linked groups will assess this problem differently based on their resources and strategic circumstances. Nonetheless, we can make broad judgments about how different types of cells will approach this question and gain insight into when a cell may opt to use CBRN methods.

Complexity

The complexity of any terrorist attack is a function of several factors: weapon, target, delivery system, simultaneity and logistics. The complexity of these factors can be measured on technological and organizational spectrums.

Weapon: Obviously, some weapons require more complex logistical operations than others: acquiring a functional nuclear weapon likely requires more planning, funding and expertise than does a simple chemical weapon that could easily be assembled with household items. Of course, most weapons fall somewhere in between the two extremes. Even different kinds of attacks employing similar weapons can vary widely in their complexity. Acquiring smallpox would be extremely difficult, whereas introducing Hoof and Mouth disease to American cattle would be an extremely simple, yet economically damaging, attack. Of course, different cells may measure different challenges associated with very similar weapons. A cell founded by a biological scientist will likely find the challenges of developing and delivering a biological weapon less onerous than a cell without such expertise.

Target: Attacking some targets is easier than others. Distributing a biological weapon in the US Capitol, though it has been done in the past, is now much more difficult than it would be in a major shopping mall. Generally speaking, targets with increased symbolic value are better defended than other potential targets. Population centers, although they likely have more security than rural areas, remain largely undefended from weapons designed to indiscriminately kill rather than explicitly target a specific physical location.

Delivery: Identical weapons can have vastly different effects depending on how they are delivered. The anthrax attacks on the US Capitol are a good example. The anthrax used in the attack would have been disastrous if the attacker had aerosolized the spores rather than delivering them through an envelope in the mail. Likewise, a nuclear weapon detonated in the air – perhaps in an airplane, which might require training a pilot – will have a much wider blast radius than one detonated in the back of a truck at ground level.

Simultaneity: Planning simultaneous attacks in varied geographic locations also requires increased sophistication. Such attacks usually require a larger organization to develop and deliver multiple weapons; and effective coordination demands more planning and, often, a skilled leadership figure that may not be necessary for a single attack.

Logistics: Acquiring, transporting and securing materials for a terror attack is often the most difficult aspect of a terrorist operation. For sensitive or unstable materials, like some biological weapons, transportation and storage may require specific technological devices. Likewise, some attacks demand organizational creativity to acquire explosives or other critical materials.

Figure 2.2 illustrates a way of assessing the complexity of a CBRN terrorist attack and locates four events within the model which exemplify the different kinds of attacks. The Aum Shinrikyo sarin gas attack in the Tokyo subway was a moderately complex attack focused on physically killing as many people as possible as a means to bring about the end of the world. The attack was only moderately complex because sarin is relatively simple to make and the delivery vehicle was an operative puncturing a plastic bag in a subway car.[13] Nonetheless, because the attack was directed and organized by the highly complex Aum organization, the plot itself was logistically complicated. The sarin was manufactured in large amounts by people other than the actual executors of the attack, all of which introduced more complexity and risk into the operation.[14] Ascribing rational purpose to a terror attack intended to initiate Armageddon is inherently difficult, but it is fair to say that Aum's purpose was to kill as many people as it could rather than simply frighten the Japanese population. Likewise, the purpose of Aum's attack was not to attract new recruits or inspire its supporters.

The anthrax attacks at the US Capitol were more technologically complex than Aum's sarin attack; the anthrax found at the Capitol site was very high quality. Only a skilled – or very lucky – technician would be able to refine anthrax to the level found at the Capitol.[15] Nonetheless, the method of delivery was very simple and formulated to warn potential victims of the danger, which suggests the attack was designed to frighten and intimidate rather than cause large numbers of casualties. Targeting a government facility in such a manner seems to indicate that the attacker was more concerned with sending a message rather than actually causing destruction.

Kamal Bourgass, who was suspected of being a member of GSPC (an al-Qaeda affiliate group), was convicted of killing a police constable in London and aiming to create a "public nuisance by the use of poisons and/or explosives."[16] Police believed that Bourgass intended to smear ricin on the door handles of cars in

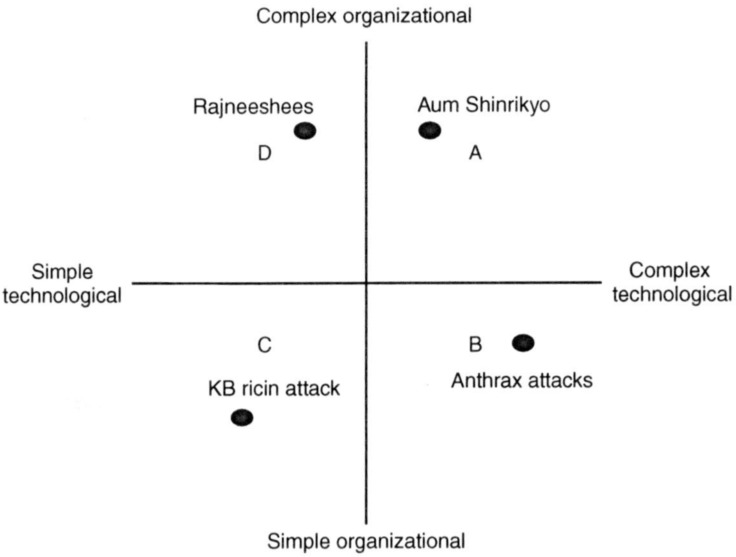

Figure 2.2 A complexity model for terrorist attacks.

London.[17] The purpose, police felt, was not to kill many people – the tactic would have been unlikely to harm anyone seriously – but to frighten people by introducing ricin into a public environment, a judgment that was reaffirmed recently in an asylum judgment regarding "W," a secondary conspirator in the "Ricin Plot."[18] The Bourgass ricin plot is an example of a CBRN attack both technologically and organizationally simple. Ricin recipes in Bourgass' lab were very simple and the delivery mechanism was reportedly no more complicated than mixing ricin with hand cream and applying it to car doors. Although Bourgass was described as a member of al-Qaeda by one of his co-conspirators, Mohammed Maguerba, all evidence suggests that he was working independently and only discussed his plans with a few close associates.[19] The simple plot seems to have been calibrated to use popular fear of biological and chemical weapons to create fear and panic disproportionate to the actual threat to the public. According to British anti-terror chief Peter Clarke, "The impact on the public, if he (Bourgass) had succeeded in what he wanted to do, is incalculable."[20]

The Rajneeshee salmonella attack illustrates another technologically simple CBRN attack. The Rajneeshees distributed salmonella in salad bars in The Dalles, Oregon in order to depress turnout in a local election. The Rajneeshees ordered their weapon from a medical supply depot and disseminated it by manually distributing on local salad bars. The purpose of the attack was not to frighten victims but to physically prevent them from taking part in the local election.

In all four of these attacks, it is important to assess the purpose of that attack from the perpetrators point of view. Why did they choose to use a CBRN weapon? The Aum sarin attack followed several previous attempts to employ chemical and biological weapons in an effort to kill large numbers of people. The delivery mechanism was simple, but the vulnerability of individuals at the target suggests that Aum hoped to kill as many people as possible. In the anthrax attacks on Capitol Hill, the attack seems to have been designed to attract as much attention as possible while minimizing casualties. The delivery of the attack was designed to maximize the perpetrators chance of avoiding detection, rather than ensuring a successful attack.

Impact

The impact of any terrorist attack is a function of the physical destructiveness of the weapon itself and the psychological impact the attack has on both the target population and the "viewing audience," particularly among potential supporters of the terrorist group. It is important to distinguish between the psychological and physical effects of an attack because an attacking group may try to alter their attack plan in order to achieve different kinds of effects.

Obviously, large attacks that kill numerous people have dramatic psychological impacts, but terrorist attacks can be designed to create psychological results without causing much destruction. Numerous groups, including ETA (Euskadi Ta Askatasuna) and the IRA (Irish Republican Army), reported the general location of their bombs before they detonated in order to minimize casualties. The purpose of such "attacks" was not physical destruction, but rather to create a psychological response in Spanish and English society. The unconventional nature of CBRN weapons – even those that do not cause extraordinary destruction – will likely increase the psychological impact of a CBRN attack on the target society. Likewise, the weapon's novelty will likely generate increased global attention.

Figure 2.3 illustrates the impact of each of the four plots described above along two axes: physical and psychological. A terrorist organization's strategic outlook on violence can be generally characterized by these two dimensions of impact. Most terrorist groups seek to circumscribe the level of destruction they cause for moral and strategic reasons – escalation of physical violence creates negative political repercussions. Conversely, millenarian groups are often solely focused on physical destruction, rather than producing psychological effects. AQ Central values the psychological impact of its attacks, but it is not afraid to produce massive casualties.

The four attacks shown in Figure 2.3 demonstrate that most terrorist attacks are limited in either their physical or psychological dimensions. Although operational factors – such as risk of detection, technological proficiency and logistics – shaped the way these operations were conceived, all of the attacks reflected the core strategic goals of the offending group. AQ Central's strategic goals are much more expansive than any of the groups reflected in Figure 2.3. Their expansive strategic goals and ideological willingness to kill means that the primary restrictions on their level of violence will be a function of resources and timing. Whereas other groups tend to conceive of operations in order to purposely limit one or both dimensions of impact, al-Qaeda's ideology suggests

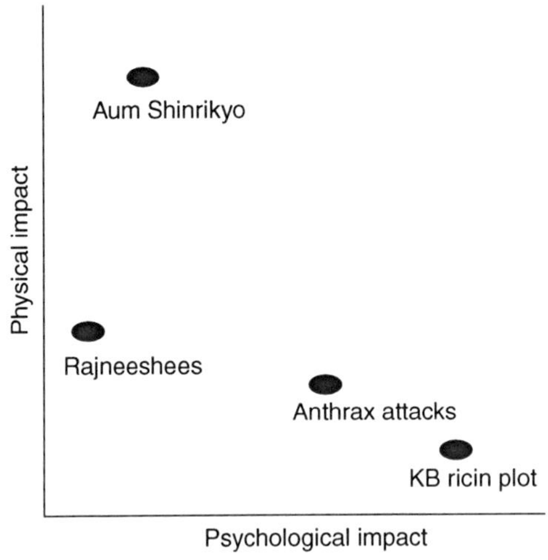

Figure 2.3 Evaluating the physical and psychological impact of terrorist attacks.

that they will aim to conceive of attacks that would maximize both physical and psychological harm.

The four dimensions of al-Qaeda and WMD

One reason that discussions about al-Qaeda's objectives often fail to resolve serious questions about the threats facing the world is that the term "al-Qaeda" is often used to refer to a variety of forms of Salafi-Jihadi militancy. Our linguistic and analytical imprecision often lumps together groups with similar ideologies, but vastly divergent strategic capabilities and circumstances. This is particularly true regarding CBRN weapons. In addition to considering the moral and strategic consequences of an attack, every al-Qaeda-linked cell considering a CBRN attack must consider the costs associated with procuring and delivering the weapon. Some attacks that would be considered feasible to some elements of "al-Qaeda" are simply beyond the logistical, financial and technical capability of other cells that support al-Qaeda's mission and ascribe to its ideology.

The framework and models described above allow us to more effectively analyze the fundamental strategic challenges and opportunities of each of the four dimensions of al-Qaeda, and then assess which of the four categories of WMD attack is likely to be most amenable to each dimension. As described earlier, each of these four dimensions bring the same basic (though there are some important differences) ideological approach to their violence, but they operate in very different strategic environments. Figure 2.4 illustrates how complex the attacks of various dimensions of al-Qaeda are likely to be. The

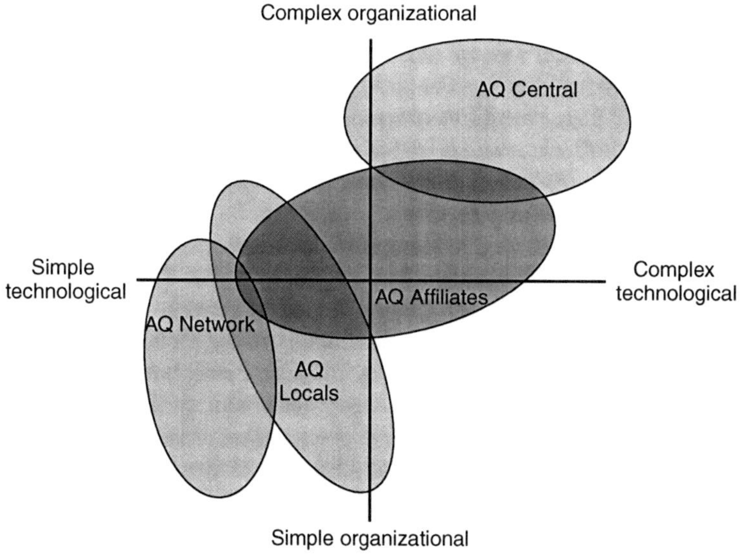

Figure 2.4 Comparative interest in WMD use by each al-Qaeda dimension.

following discussion examines each of these in-depth. Obviously such judgments are broad and inevitably flawed to some degree, but they represent a best effort to understand how various al-Qaeda-supporting groups will assess the costs and benefits of various kinds of attacks.

AQ Central

AQ Central demonstrated its willingness to kill thousands of people simultaneously on September 11, 2001. That operation required moderate technical skill – the ability to steer an airplane – and an extensive logistical and coordination capability that seems far out of reach for AQ Locals or the AQ Network. AQ Central must make a complex strategic calculation to determine whether the benefit from developing and deploying CBRN weapons is worth the risk and cost associated with them. AQ Central has advanced three operational missions for itself: first, inspiring the community of Muslims (Ummah) to join their movement; second, dragging the West into fights where it can be attacked and confronted over the long-term; third, bleeding the United States economically to weaken it in the long-term.[21] The question is whether – and how – employing WMD will help AQ Central achieve these missions.

Importantly, AQ Central has focused on attacking the "far enemy" – primarily identified as the United States, but including other Western democracies as well. Part of the rationale for this decision seems to be that AQ Central is loathe to target Muslims for fear of losing its reputation as the vanguard of Muslim resistance to encroaching Western cultural and political imperialism. Although AQ Central's ultimate goal is to destroy apostate governments in the Middle East, its leaders understand that killing Muslims is counterproductive to that goal.

The West faces a significant challenge deterring AQ Central from using WMD, largely because of AQ Central's geographic distribution and its tendency to embed itself within host societies full of innocents. AQ Central certainly understands the difficulties the United States would have retaliating against a WMD attack, but that does not mean AQ Central can ignore the possibility of American retaliation. Indeed, a variety of al-Qaeda-linked groups have used attacks in order to provoke misguided retaliation from the target society.

In the past, AQ Central has tried to manipulate such responses, but experience has shown that to be more difficult than they originally hoped. The 9/11 attacks illustrate this strategic dilemma. AQ Central hoped to provoke an American invasion of Afghanistan that would be as detrimental to the United States as it was to the Soviet Union 20 years before. That invasion came, but it was small, carefully calibrated and – in the short-term – very successful. In the wake of the US response to 9/11, a number of AQ Central commanders protested vigorously against further provocation of the United States because of the detrimental effect the US retaliation had on al-Qaeda operations.[22]

Indeed, AQ Central is unlikely to speciously use CBRN. Like all sophisticated terrorist organizations, AQ Central understands that timing and targeting are important to the success of an attack and that mass killing alone is unlikely to produce the political effects they desire.[23] Furthermore, AQ Central has already demonstrated a cautious approach to WMD, likely because it understands the potential costs of such an operation. Even a relatively small chemical attack on an American city would almost certainly provoke a large American response of some sort; AQ Central will want to be sure the benefits of its attack outweigh the costs.

Reportedly, Ayman al-Zawahiri canceled a hydrogen cyanide attack on the New York subway system that had been planned by al-Qaeda member Emir Yusuf al-Ayiri in Saudi Arabia.[24] That decision illustrates the strategic questions AQ Central must resolve regarding CBRN use. The minimal details available about the al-Ayiri attack suggest casualties could easily have been in the hundreds, or even thousands, but that the attack would not have debilitated New York city the way a nuclear or large-scale biological attack would. According to journalist Ron Suskind's account, al-Zawahiri called off al-Ayiri's chemical attack, but promised something "bigger." This tale reinforces the long-standing assumption that AQ Central's commanders feel compelled to increase the scope of their violence to demonstrate an increasing capability to potential supporters. Apparently, al-Zawahiri believed that AQ Central will only get a few shots at the US homeland and must make the most of each.

Although AQ Central likely understands the psychological impact using CBRN weapons would have on a target society, their primary audience is not the victim society, but the hearts and minds of Muslims around the world. A chemical attack in the New York subway would certainly terrify an American population, but might not provide appropriate visible evidence for al-Qaeda supporters to justify the inevitable retaliation from such a deadly and frightening attack.

AQ Central's intention to maintain its position as the vanguard of the Ummah is critically important. Previous AQ Central attacks have been simultaneous attacks on symbolic and hardened targets. Such attacks require increased complexity because of the intelligence challenge of inserting attackers and weapons onto hardened targets and the need to understand the tactical environment around multiple targets. Taking on and meeting such challenges is an end in itself for AQ Central because the group wants to demonstrate to potential supporters that it is capable of meeting such challenges and serving as the vanguard of Muslims.

AQ Central would likely not risk the inevitable retaliation from a WMD attack unless it could credibly be expected to garner massive attention around the Arab and Muslim world. Since AQ Central's commanders seem to believe that its credibility with potential supporters in the Ummah is partly a function of its ability to demonstrate the capacity to complete sophisticated attacks, we should expect that AQ Central attacks of all kinds, including CBRN, will be more complex. Although AQ Central would certainly be amenable to simple CBRN tactics, such as destroying chemical facilities, they are likely to complicate such a mission by attempting to strike multiple targets simultaneously. Although that strategy might not increase the psychological impact of the attacks on the target society, they demonstrate AQ Central's sophistication and overall capability to its supporters among the Ummah.

AQ Affiliates

The most important AQ Affiliates operating in the world today are al-Qaeda in Iraq (AQI) and al-Qaeda in the Islamic Maghreb (AQIM). AQI is the most active terrorist organization in the world and, because of global interest in Iraq and AQI's sophisticated media operations, is also the most visible. AQIM's operations receive far less attention globally than AQI, but AQIM has increased its activities and is thought to have a strong network of operatives and supporters in Europe. AQIM's penetration into Europe is important because many AQ Affiliate cells would likely prefer to use CBRN weapons against Western targets rather than in the Middle East.

AQ Affiliates operating in the heart of the Middle East often have a less globalized perspective than AQ Central. Even after Abu Mus'ab al-Zarqawi joined al-Qaeda in October 2004, he was focused on the "near enemy" rather than the "far enemy" targeted by AQ Central.[25] Although most AQ Affiliates do not flout AQ Central's strategic doctrine as bluntly as al-Zarqawi did, AQ Affiliates tend to focus their attacks at local and regional figures that present a less appetizing CBRN target than major Western cities.

Reports accusing al-Zarqawi of planning a chemical attack in Amman, Jordan highlights one potential AQ Affiliate approach to CBRN. Al-Zarqawi planned for five truck-borne bombs to target several government targets around Amman. Analysts suggest that the attack, if successful, would have released a poison cloud that could have killed up to 20,000 people.[26] Despite such drastic estimates, the attack was relatively simple. The chemicals were primarily designed to enhance the explosion itself and were to be dispersed mainly by the force of an explosion rather than a dedicated dissemination device. The Jordanian government claimed that al-Zarqawi had planned to use chemical weapons in the attack – a charge that al-Zarqawi later claimed was designed to discredit him.[27] Whether or not the Jordanian government's claims were accurate, al-Zarqawi's unwillingness to publicly embrace the chemical weapons charge illustrates the strategic paradox such weapons offer an AQ Affiliate. CBRN weapons are so frightening and novel that even a terrorist as brutal as al-Zarqawi feared the public backlash he expected from using such weapons on a Muslim country.

Al-Zarqawi's "chemical" plot demonstrated at least some organizational complexity – money for the attack was imported from Iraq and the delivery of five truck bombs simultaneously requires a strong organizational structure.[28] Furthermore, the use of chemical accelerants for the explosives – even if not conceived as a chemical weapon – may have had a similar psychological effect on the Jordanian population as a poorly distributed chemical weapon.

Chemical attacks inside Iraq also provide a model for AQ Affiliate CBRN warfare. Although these attacks have successfully killed numerous people, the kinetic blast rather than latent chemicals killed most victims. Such attacks have been relatively simple, mostly just large improvised explosive devices attached to a chemical tanker truck.[29] The unconventional element seems primarily designed to attract attention to the attacks and inspire fear in the target population.

Attacks by AQ Affiliates using CBRN are likely to vary widely. Large CBRN attacks will likely be aimed at Western targets because such attacks on largely Muslim targets would likely produce unhelpful backlash in the local population. Conversely, AQ Affiliates that use CBRN weapons in the scope of their local conflict are likely to focus on simpler attacks like those used by AQI. The need to moderate violence to avoid a dramatic backlash from the local population and the pressures of continued tactical operations will encourage AQ Affiliates to aim for simpler operations designed to enhance the impact of conventional attacks.

AQ Locals

Although cells in both AQ Central and AQ Locals have succeeded in pulling off deadly attacks against civilians, such cells are just as notable for their failures – see the recent propane bombs in London and at Glasgow airport – as their successes. A cell of AQ Locals will generally have fewer resources and far less expertise than either AQ Central or AQ Affiliates. AQ Locals tend to adopt AQ Central's globalist perspective, though recent attacks have been linked to the war in Iraq.[30] Nonetheless, there is some debate among ex-patriot Salafi-Jihadi scholars over the theological acceptability of attacking Western states. Abu Basir al-Tartusi, an influential Syrian preacher living in London, has argued against both suicide bombings and the acceptability of Salafi-Jihadi in Britain attacking their British "hosts."[31]

A small cell of AQ Locals may have some advantages over their more organized cousins. For example, individuals living within a target society may be able to identify targets more easily and almost certainly have a better understanding of how to frighten and intimidate the local population. To be successful, this kind of cell will almost certainly avoid complex operations, focusing instead on low-tech, logistically direct operations designed to create chaos and fear. From a strategic standpoint, a cell of AQ Locals embedded within its target society stands little chance of remaining operational after a major CBRN attack, although it might be able to wage a low-level campaign using discreet chemical or biological weapons designed to inspire fear over a sustained period of time. Although the 7/7 bombers in London certainly demonstrated that AQ Locals are willing to die to commit attacks, their links with the failed 7/21 bombers suggests the 7/7 attack was designed as the beginning of a campaign.[32]

Kamal Bourgass' alleged plot to frighten the citizens of London with ricin is an excellent example of a CBRN plot by an AQ Local cell. Bourgass had a history of operating in militant organizations, but was not directly coordinating with or taking orders from more organized elements of al-Qaeda. He understood his technical and logistical limitations and, if London police are to be believed, determined that the best way to have a major impact was an operation designed to frighten people rather than cause destruction.

A CBRN attack by a cell of AQ Locals will likely be very simple and designed to use the comparative advantage of such cells – extensive knowledge of the target society, both physically and culturally. The limited resources and tactical expertise available to AQ Locals represents an inherent limitation on their ability to develop a complex, massive physical attack. Cells with limited means to operationalize destructive conventional attacks may decide that simple, small-scale chemical and biological attacks – though no more dangerous than their conventional counterparts – offer a way to maximize the political impact of the cells' limited capabilities.

AQ Network

Al-Qaeda supporters without ties to any militant organization will have limited operational choices because they lack tactical training and logistical support. Rather than being driven by a desire to optimize their strategic impact while minimizing the strategic costs associated with an attack, an AQ Network cell will probably take advantage of whatever resources it finds available. The strategic dynamics of an AQ Network cell closely resemble that of AQ Locals, except that its tactical capacity is probably even lower. The driving strategic logic behind a successful CBRN attack by an AQ Network cell is that it is simple.

Whereas AQ Central, AQ Affiliates, and even some AQ Locals may complicate a plan in order to enhance its physical or psychological impact, AQ Network cells will likely minimize complexity at all costs. That does not necessarily mean that the threat of a CBRN attack by an AQ Network cell is low or that such an attack would not be devastating. A chemical engineer that supports al-Qaeda for ideological reasons would be extremely dangerous, and potentially very deadly. Likewise, unprotected infrastructure that could be attacked with simple, conventional tactics but produce an unconventional outcome – such as a chlorine rail car – would be prime targets for an AQ Network cell. Furthermore, an AQ Network cell relatively unencumbered by pressure to lead the Ummah may be more willing to attack less symbolic targets, or at least less visibly symbolic targets. Such groups are more likely to target the water supply of an apartment building, for example, than take on the harder task of poisoning water supplies of a critical government facility. AQ Central, cognizant of the role it hopes to play as vanguard of Ummah, is unlikely to invest in such an unspectacular sort of attack unless it is on a very large scale.

Conclusions

Frightening people en masse is easier than killing them. It is no surprise that AQ Central is more likely than its less organized cousins to attempt complex and destructive attacks using WMD. Expectations of AQ Central among its supporters are higher than those of other al-Qaeda dimensions; furthermore AQ Central has extensive logistical capabilities that enable it to more reasonably attempt complex operations. Conversely, AQ Network cells have minimal logistical and tactical capabilities, but may actually have better cultural and political intelligence than their more organized counterparts. The greatest threat from CBRN among members of this al-Qaeda dimension would be simple attacks on poorly protected industrial infrastructure that will produce catastrophic results as a function of chemical, biological or radiological releases.

From this analysis, we conclude that Western governments can take a variety of steps to prevent an al-Qaeda-linked CBRN attack.

1 *Emphasize norms against CBRN use.* Normative prohibitions on CBRN function even in an al-Qaeda context, albeit via a different mechanism than in traditional state relationships. Al-Zarqawi was clearly concerned that he would be blamed for a chemical attack, and feared a backlash from an appalled Jordanian public. Western governments should work to ensure that terrorist leaders will fear a commensurate backlash if a CBRN attack is attempted in the West. Highlighting "attacks" like that in Amman, along with educational campaigns worldwide – especially in societies that al-Qaeda depends on for support and recruitment – can help raise the strategic and operational costs for the use of CBRN by centralized al-Qaeda groups.
2 *Develop informal monitoring networks.* Governments cannot defend against CBRN by themselves. Informal professional and social organizations with expertise in components of CBRN weapons are critical. Biologists, chemists and physicists need to keep tabs on their own. The first warning signs that something is amiss may come when scientific community members pull back from their professional colleagues – declining to attend conferences, losing interest in research, etc.
3 *Educate the Public.* Common-sense safeguards against CBRN attacks are critically necessary, but ultimately an attack may occur. Planning to prevent a CBRN attack should be integrated with efforts to mitigate their effect. To a large degree, that means public education efforts are needed, not just to educate people about how to behave in the event of an attack, but also to limit the psychological impact of an attack on society. The UK government's emphasis on fostering social "resilience" in their homeland security policy and operations is particularly instructive here.

If there is any lesson to draw from al-Qaeda post-9/11, it is that al-Qaeda in all of its forms will find ways to be unpredictable. Despite the organization's inherent volatility, however, we should be able to make broad prognostications about the kinds of attacks to expect from al-Qaeda. Al-Qaeda – in all of its formulations – tries to optimize its political impact. Its ideology prescribes brutal violence on a massive scale, but it is not pointless or indiscriminate. AQ Central, in particular, rationally assesses the costs and benefits of every operation it conducts. Smart terrorist groups understand that they can undermine their own cause with misplaced, counterproductive violence. Less professional al-Qaeda cells may not make such careful judgments. One only need assess al-Zarqawi's dramatic mistakes as the Emir of AQI to understand that not all deadly terrorists are politically sophisticated. Indeed, the fact that al-Qaeda's sophisticated center is politically linked to its less professional affiliates may be one of its most important strategic weaknesses. Exploiting the strategic miscalculation by amateurs may be one of our best weapons against al-Qaeda.

Acknowledgments

The views expressed herein are those of the authors and do not purport to reflect the position of the United States Military Academy, the Department of the Army, the Department of Defense, or the US government.

Notes

1 Michael Barkun, "Terrorism and Doomsday," in James J.F. Forest (ed.) *The Making of a Terrorist, Volume 3: Root Causes* (Westport, CT: Praeger Security International, 2005).
2 John Parachini, "The Making of Aum Shinrikyo's Chemical Weapons Program," in James J.F. Forest (ed.) *The Making of a Terrorist, Volume 2: Training* (Westport, CT: Praeger Security International, 2005) pp. 277–95. Also, for a list of Aum attacks with biological agents, see David E. Kaplan, "Aum Shinrikyo," in Jonathan B. Tucker (ed.) *Toxic Terror: Assessing Terrorist Use of Chemical and Biological Weapons* (Cambridge, MA.: MIT Press, 2000), p. 221.
3 Bruce Hoffman, "Challenges for the U.S. Special Operations Command Posed by the Global Terrorist Threat: Al Qaeda on the Run or on the March?" *Testimony for House Armed Services Subcommittee on Terrorism, Unconventional Threats and Capabilities*, February 14, 2007.
4 This description of Islam and al-Qaeda's ideology is courtesy of William McCants, a Fellow at the Combating Terrorism Center at West Point.
5 For a complete analysis of the Salafi-Jihadi ideology, please see William McCants and Jarret Brachman, *The Militant Ideology Atlas*, West Point, NY: Combating Terrorism Center, 2006. Online at: http://ctc.usma.edu.
6 Reuven Paz, "Global Jihad and WMD: Between Martyrdom and Mass Destruction," in Hillel Fradkin, Husain Haqqani and Eric Brown (eds.) *Current Trends in Islamist Ideology (Volume 2)* (Washington, DC: Hudson Institute, 2005).
7 Suleiman Abu Gheith "In the Shadow of the Lances," (June, 2002). Selected translation online at: www.memri.org/bin/articles.cgi?ID=SP38802.
8 Quoted in Robert Wesley, "Al-Qa'ida's WMD Strategy After U.S. Intervention in Afghanistan," *Terrorism Monitor*, 3:20.
9 Abu Hamzah Al-Muhajir (Abu Ayyub al-Masri), *The Mujahidin News*, audio statement, September 28, 2006.
10 BBC News, "Man Admits UK–US Terror Bomb Plot," *BBC News Online*, October 12, 2006. Online at: http://news.bbc.co.uk/1/hi/uk/6044938.stm.
11 Reuven Paz, "Gobal Jihad and WMD," p. 82.
12 See www.ctc.usma.edu/harmony_docs.asp, "al-Adl Letter." The letter is written by "Abd-al-Halim-Adl" who may be Sayf al-Adl, AQ Central's former military chief.
13 For a complete account of this attack, please see James M. Smith, "Japan and Aum Shinrikyo," in by James J.F. Forest (ed.) *Countering Terrorism and Insurgency in the 21st Century,* Vol. 3 (Westport, CT: Praeger, 2007), pp. 549–68.
14 John Parachini, "The Making of Aum Shinrikyo's Chemical Weapons Program," pp. 277–95.
15 David Johnston and Alison Mitchell "A Nation Challenged: The Widening Inquiry, Anthrax Mailed to Senate is Found to be Potent Form; Case Tied to Illness at NBC," *New York Times*, October 17, 2001.
16 Bruce Hoffman, "CBRN Terrorism Post-9/11," in Russell Howard and James J.F. Forest (ed.) *Terrorism and Weapons of Mass Destruction* (New York: McGraw-Hill, 2007). This chapter can also be found online at: http://fletcher.tufts.edu/jebsencenter/researchbriefs/JCCTS_Hoffman_CBRN_01–2007.pdf.
17 Ibid.

18 Ibid. Justice Mitting in the matter of "W" and the Secretary of State for the Home Department, May 14, 2007. See: www.icj.org/IMG/SIAC-W.pdf.
19 Vikram Dodd, "Doubts Grow over Al-Qa'ida Link in Ricin Plot," *Guardian*, April 16, 2005. Online at: www.guardian.co.uk/terrorism/story/0,12780,1461030,00.html.
20 BBC News "Killer Jailed Over Poison Plot," *BBC News Online*, April 13, 2005. Online at: http://news.bbc.co.uk/1/hi/uk/4433709.stm.
21 One of the best recent statements on al-Qaeda strategy is: Yaman Mukhaddab, "Al-Qa'ida Between a Past Stage and One Announced by Al-Muhajir," *Keepers of the Promise* website (accessed November 18, 2006). Mukhaddab describes eight goals that have been "achieved" by Al-Qaeda since the attacks of 9/11. They can be summarized as above.
22 Vahid Brown, "Cracks in the Foundation: Leadership Schisms in Al-Qa'ida 1989–2006," The Combating Terrorism Center, September 2007. Online at: www.ctc.usma.edu/aq/pdf/Harmony_3_Schism.pdf.
23 Magnus Ranstorp, "Statement to the National Commission of the Terrorist Attacks Upon the United States," March 31, 2003. Online at: www.globalsecurity.org/security/library/congress/9–11_commission/030331-ranstorp.htm.
24 Ron Suskind, *The One Percent Doctrinrk* (New York: Simon & Schuster, 2006) pp. 218–20.
25 Brian Fishman, "After Zarqawi: The Dilemmas and Future of Al-Qa'ida in Iraq," *Washington Quarterly*, Autumn 2006.
26 Rohan Gunaratna, "Abu Musab al-Zarqawi: A New Generation of Terrorist Leader," *IDSS Commentaries*, July 5, 2004.
27 BBC News "Al-Qaeda Denies Jordan WMD Plot," *BBC News Online*, April 30, 2004. Online at: http://news.bbc.co.uk/2/hi/middle_east/3672891.stm.
28 Alfred Prados and Jeremy Sharp "Jordan: U.S. Relations and Bilateral Issues," Congressional Research Service, July 14, 2006.
29 Damien Cave, "Iraq Insurgents Employ Chlorine in Bomb Attacks," *New York Times*, February 22, 2007.
30 Victoria Burnett and Alissa Rubin "Doctor Accused in Glasgow Attack Described as Angry About the Iraq War," *New York Times*, July 5, 2007.
31 Mohammed Al-Shafey, "London Based Salafi Scholar Issues Fatwa Prohibiting Suicide Operations," *Asahrq al-Awsat*, August 27, 2005; Abu-Bashir al-Tartusi, "Statement on the Recent Events in the UK," August 15, 2005. Online at: www.e-prism.org or www.abubaseer.bizland.com/hadath/Read/hadath20.doc.
32 Christopher Dickey, "Outward Bound?" *Newsweek*, July 28, 2005; Mitchell Silber and Arvin Bhatt, "Radicalization in the West: The Homegrown Threat," New York Police Department report.

3 Al-Qaeda's thinking on CBRN

A case study

Anne Stenersen

Introduction

In May 2003, the radical Islamist cleric Nasir bin Hamd al-Fahd issued a fatwa legitimizing the use of weapons of mass destruction (WMD) against the United States.[1] In a frequently quoted passage he states:

> The attack against it [the United States] by WMD is accepted, since Allah said: "If you are attacked you should attack your aggressor by identical force." Whoever looks at the American aggression against the Muslims and their lands in recent decades concludes that it is permissible .... Some brothers have totalled the number of Muslims killed directly or indirectly by their weapons and come up with a figure of nearly 10 millions.[2]

Shaykh al-Fahd's fatwa is often used to illustrate that the al-Qaeda network has a clear intention of carrying out a chemical, biological, radiological or nuclear (CBRN) attack in the West. Few studies, however, have attempted to interpret the fatwa in relation to what is actually taking place within the global jihadi current today. In the hype surrounding al-Qaeda's alleged pursuit for CBRN weapons, we tend to overlook the fact that no terrorist attack involving CBRN materials has actually been carried out by al-Qaeda so far. Al-Qaeda's interest in deploying a CBRN weapon is, perhaps, less than anticipated.

The aim of this study is to fill a gap in current research on al-Qaeda's CBRN capabilities. Although there is a large literature on al-Qaeda and CBRN terrorism, few studies have actually subjected al-Qaeda's own documents and literature on CBRN weapons to critical scrutiny.[3] While such literature has traditionally been difficult to access, al-Qaeda's active use of the internet has over recent years given researchers easy access to a range of primary sources about militant Islamism, including a large collection of jihadi training manuals and handbooks.

This study is based on CBRN manuals and discussions found on radical Islamist web pages affiliated with the al-Qaeda network. This chapter will provide an overview of the nature of these manuals, as well as analyzing them in a broader context. How have these manuals developed? To what extent are they being revised and discussed? And, ultimately, what can they tell us about the nature of al-Qaeda's interest in CBRN weapons?

The study reveals two important findings: first, that the subject of CBRN weapons receives very limited attention among al-Qaeda's followers online; second, that the online manuals and discussions display a lack of knowledge and innovative ability regarding CBRN-related means and methods on the part of jihadists.

Al-Qaeda's use of the internet

It is well known that today the internet plays a vital role for al-Qaeda and its loose network of affiliates and sympathizers. Most importantly, the internet is utilized for propaganda and communication purposes, but is also used for training, recruitment, fund-raising and electronic jihad.[4] After the disappearance of the training camps in Afghanistan, it has become popular to view the internet as a "virtual Afghanistan,"[5] referring to the numerous training manuals and handbooks available online, including high-quality instruction videos. Many of the jihadi discussion forums have specialized sub-forums in which members exchange information and experiences on a range of military and technical subjects. However, the subject area of training remains relatively small. In one of the largest discussion forums, *al-Firdaws*, which at the time of access (December 19, 2006) contained some 19,000 threads, 750 threads (or 3.9 percent) were posted in the "equipment and preparation" sub-forum (*muntada al-'udda wal-i'dad*). In comparison, 24 percent of the threads were posted in the sub-forum for communiqués and reports (*muntada al-bayanat wal-taqarir*). A survey of other forums showed a similar pattern of distribution.[6] That being said, there is still a considerable amount of training material and a number of manuals available online, as well as daily activity in the sub-forums dedicated to this topic. A well-known online collection of training manuals is the *Encyclopedia of Preparation* (*mawsu'at al-i'dad*), which in 2004 contained around 300 Arabic-language documents amounting to some 10,000 pages, in addition to a number of English-language documents and handbooks. The collection is compiled from a range of sources: some documents appear to be written by veterans of the Afghan war or other experienced jihadists, while other manuals are simply translations from English literature such as *The Terrorist's Handbook* or US Army Field Manuals. The *Encyclopaedia* has been updated several times, and is currently in its fourth edition.[7]

The online popularity of CBRN weapons

It is important to emphasize that CBRN-related discussions and material constitute a very small part of the large jihadi literature and discourse found online. At the time of writing, around ten CBRN manuals had been identified (see Table 3.1), in addition to one English-language video on how to extract ricin.[8] In comparison, there are around 40 Arabic-language instruction videos

Table 3.1 A list of CBRN manuals found on jihadi web pages

Title	*Original title*	*Type*	*Format*	*Pages*
The Making of Poisons	*sina'at al-sumum*	BC	Pdf	8
The Mujahideen Poisons Handbook	*The Mujahideen Poisons Handbook*	BC	Pdf	19
Course in Homemade Poisons and Poisonous Gases	*dawrat al-sumum wal-ghazat al-samma al-sha'biyya*	BC	Pdf	30
The Poisons Encyclopaedia/ Comprehensive Course in Poisons and Poisonous Gases[a]	*mawsu'at al-sumum/ dawrat al-sumum wal-ghazat al-samma al-kubra*	BC	Pdf	99
Poison Warfare	*harb al-sumum*	BC	HTML	10
The Unique Invention	*al-mubtakar al-farid*	C	Pdf	12
Biological Weapons	*al-silah al-biyuluji*	B	word	19
Preparation of Botulism Toxin (rotten food poison)	*tahdir samm al-butulizm (samm al-ta'am al-fasid)*	B	Pdf	28
Radiological Pollution	*al-talawwuth al-ish'a'i*	R	HTML	2
The Nuclear Bomb of Jihad and How to Enrich Uranium	*al-qunbula al-nawawiyya al-jihadiyya wa kayfiyyat al-takhsib al-nawawi*	N	Pdf	479

Note

a This manual has been distributed in at least two versions: a HTML document entitled *The Poisons Encyclopaedia* and a Pdf document entitled *Comprehensive Course in Poisons and Poisonous Gases*. The content of the two versions is otherwise identical except that images are left out of the HTML version.

available on conventional weapons and how to make various types of explosives.[9] The total amount of *written* training material is hard to estimate, but the contents in the previously mentioned *Encyclopaedia of Preparation* might give an indication: in 2004 the collection contained some 300 documents on a range of topics including conventional warfare, improvised weapons, guerrilla tactics and physical training. Although the original homepage of the *Encyclopaedia* is not active today, the collection is widely circulated on jihadi discussion forums, in addition to an unknown amount of new training material that has been produced since 2004. It seems reasonable to assume, therefore, that the ten CBRN manuals identified constitute merely a few percent of all training material available on al-Qaeda-affiliated web pages.

Another indicator of the online popularity of CBRN weapons is to look at the topics most frequently discussed in jihadi discussion forums. An analysis of *al-Firdaws* revealed that out of 764 threads posted in the "equipment and preparation" sub-forum, only 35 (4.6 percent) were CBRN-related.[10] In comparison, 42 percent of the threads were related to various types of explosives. The analysis also showed no indication that the topic of CBRN is becoming more popular, or that new handbooks or recipes are being developed. In 2005, 20 CBRN-related threads (4.7 percent of the total) were posted, while in 2006, 15 CBRN-related threads were posted (4.4 percent of the total). These threads included discussions on how to make and deliver different CBRN materials, copies of newspaper articles on al-Qaeda and CBRN and links to CBRN manuals or excerpts from such manuals.

When put into perspective, therefore, the subject of CBRN weapons seems to receive very limited attention among al-Qaeda's followers online. The rest of this chapter will examine the contents of these online sources in more detail. What is the nature of these CBRN manuals and discussions, and what can they tell us about the technical knowledge and innovative abilities of the jihadists?

Chemical and biological manuals

Eight of the ten manuals examined are related to chemical or biological materials. They can roughly be divided into three categories:

1. collections of crude poison-making methods, similar to those found in *The Poisoner's Handbook*;
2. Recipes requiring advanced laboratory equipment and skills, based on scientific articles and college textbooks;
3. a recipe for a crude, chemical device, including a fully developed dispersal mechanism.

The first category of manuals contains crude recipes of poisons and poisonous gases resembling those found in English-language literature such as *The Poisoner's Handbook*.[11] Several of the manuals also contain descriptions of experiments, indicating that the various recipes have been tested and experimented with by the jihadists themselves. The author of *The Mujahideen Poisons Handbook*, for example, states in the introduction that the manual is based on a "poisons course" that he himself had attended.[12]

The recipes described in these manuals are generally very crude, and can be carried out in an improvised facility such as a kitchen or a garage. Also, the recipes are often based on materials that are relatively easy to obtain, such as castor beans, nicotine, hydrochloric acid, potassium cyanide and so on. The recipes on how to extract ricin from castor beans, found in several of the manuals, are illustrative of the technical quality of these manuals. The recipe is essentially similar to the procedure described in *The Poisoner's Handbook*, and is based on rinsing castor bean pulp with acetone in order to remove the oil from the pulp. After drying the pulp, a white powder is obtained, which is described as "the poison" and sometimes as "pure ricin."[13] In reality, the result is far from pure. A Spanish laboratory which tested the above-mentioned recipe obtained extracts containing 0.01–0.33 percent ricin, depending on the variety of the castor plant the seed was taken from (the jihadi manuals do not specify which variety should be used). The Spanish test concluded that the low content of ricin makes the agent unsuitable as a "weapon of mass destruction," although it might be used against "selective targets, limited to one or several persons."[14]

The most comprehensive manual in this category is *The Poisons Encyclopaedia*, a 100-page manual describing some 40 different chemical agents, including

15 gases. Around 20 of these agents appear to have been tested on rabbits, and lab reports are included in the manual. The descriptions of the experiments typically consist of how much poison was given to the rabbit, how it was given (injected, given orally or mixed with an oily liquid and applied to the skin) and time until death. Some also include a description of the symptoms and whether the rabbit was "strong" or "small." The rabbit's death is then used as "proof" that the recipes work. This testing method is not very accurate, however. The symptoms are often vaguely described, and with a few exceptions, placebo tests are not conducted. Thus, we do not know whether the rabbit died from the poison, from the solvent or from other substances manufactured instead of the intended poison. The following excerpt may serve as an illustration. Here, the author describes an experiment in which a substance thought to be botulinum toxin is tested on a rabbit. The procedure for making the substance is well-known from English literature; it consists of mixing meat with droppings or soil, putting the ingredients in a jar, filling it with water and closing it tightly, and leaving it in a tempered place for some days.[15] After providing a description of the manufacture procedure, the manual states:

> After ten days, a coffee-brown piece appeared on the surface of the water and on the glass walls. Then I took this brown substance and dissolved a small amount of it, about 0.1 g or less, in ethyl alcohol (about 5 ml). We took 1 ml of this solution and injected it into a strongly built rabbit, and it died eighteen (18) hours after the appearance of the above-mentioned symptoms.[16]

The claim that the "brown substance" manufactured in this procedure equals botulinum toxin is of course a myth. In reality the experiment is likely to yield a host of different bacteria, many of which may be capable of killing a rabbit. Even if the experiment was carried out the way it is described, the rabbit's death is no proof that botulinum toxin was produced, or illustrative of how potent the manufactured "poison" is. The description of the experiments may, on the other hand, serve to convince an unskilled reader of the recipe's validity.

Several gases, like chlorine and hydrogen cyanide, are also tested on rabbits. These experiments are conducted by leaving the rabbit in a small, confined area and releasing the gas by simply mixing two reactants. The manuals do not contain detailed information on how to use such poisons or gases on a larger scale, and the purpose of the experiment seems to have been merely to familiarize oneself with the gases and their effects. In general, the manuals in this category typically lack instructions on how to effectively deliver the agents. Delivery methods are sometimes suggested, but no dispersal device has been developed.

The second category manuals describe procedures requiring advanced laboratory equipment and skills. The two biological manuals listed above fall into this category, and they describe how to grow the bacteria *Yersinia pestis* (the cause of the plague) and *Clostridium botulinum* (producing botulinum toxin). The manuals are typically copied from scientific articles and college textbooks, rather than tested and developed by the author. The following quote, taken from the introduction of one of these manuals, illustrates this point:

> In preparing this report, I relied on a study on the theoretical characteristics of the toxin and the various methods to purify it, and I tried to pick the method which was easiest and most effective, but at the same time most inexpensive. However, I was unable to implement it in practice, because I lacked the opportunities to do so.[17]

These manuals also lack effective instructions on how to deliver the agents. There is generally little interest in these manuals, because they seem to be too advanced for the average user of the jihadi internet forums.

The third category of manuals differs from the two others as it is the only type of manual that actually describes a realistic delivery mechanism for a chemical or biological agent. Only one manual of this category has been found, entitled *The Unique Invention* (*al-mubtakar al-farid*). The manual describes a crude device for dispersing chlorine and hydrogen cyanide gas. Figure 3.1 is taken from the manual, showing the device from above.[18] It consists of a metal box containing four glass bottles filled with acid, surrounded by two chemicals. The idea is to break the glass bottles with a small explosive charge, causing the acid to mix and react with the two other chemicals and produce chlorine and hydrogen cyanide gas, which will be released through holes in the device. The manual contains detailed explanations and illustrations on how to assemble it, as well as suggestions on what kind of places to target and from where to obtain the chemicals. However, in spite of its relative sophistication, the manual does not appear to be more popular or more discussed than other online CBRN material. No reports on tests or experiments using the device have been found, although the author states in the manual that "the Invention, which is presented on these pages, is the result of experiences from the months and years of preparations and planning."[19] The document is nevertheless interesting because it is the only CBRN manual among those examined that reflects a comprehensive attempt at designing an improvised chemical device for use against an indiscriminate target. One should keep in

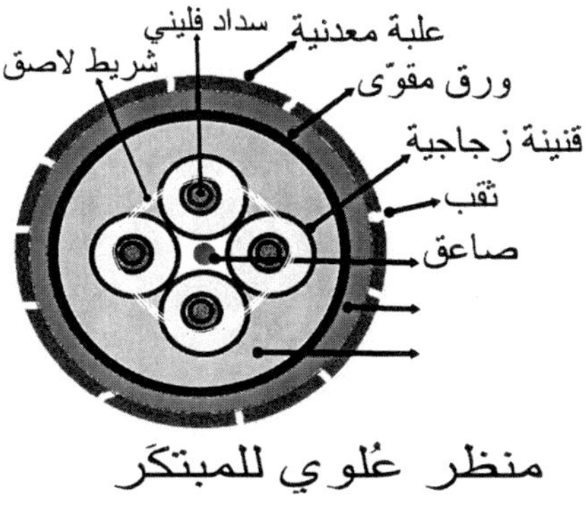

Figure 3.1 An illustration of the crude dispersal device described in *The Unique Invention*.

mind, however, that the recipe does have certain technical shortcomings, and that the actual effectiveness of the device is debatable. Another point worth noting is that there are still no known attempts by jihadists to assemble or use the device, although the manual has been available online since at least 2004.[20]

Radiological and nuclear manuals

Nuclear weapons seem to be a quite popular topic on jihadi discussion forums. A 19-lesson manual entitled *The Nuclear Bomb of Jihad and How to Enrich Uranium* appeared for the first time around 2005,[21] and since that time it has been circulated on various forums. In January 2007 it had status as a "sticky" link on the *al-Firdaws* sub-forum for "equipment and preparation," and it was the most visited thread in that sub-forum, with more than 13,000 hits. In comparison, the second most popular link at that time, entitled "Rockets (very, very important)" had around 2,600 hits.

The manual is authored by a person with the nickname "No1," who admits that he "spent two years studying nuclear physics" on the internet.[22] The manual itself confirms this picture. It consists of a collection of texts, illustrations and articles from various sources, that seem to be randomly put together, without much regard as to whether the information is correct or not. One example of the numerous technical errors in this manual is found in lessons 10–12 where the author claims that melting exactly 80.1 kg of radium with a mixture of iron oxide and aluminum (so-called "thermite") will cause a nuclear explosion similar to the Hiroshima bomb. More realistic options such as merging radium with regular explosives in order to manufacture a radiological dispersal device (RDD) are not mentioned. The chapter on how to obtain radioactive material is equally far-fetched. Lesson 18 of the manual describes how to extract uranium and other radioactive substances from black sand.[23] Any thoughts about how to obtain the material illegally on the black market are not offered. To be sure, the manual does provide a general introduction to nuclear physics and the history of the nuclear bomb, but not much more.

It has sometimes been reported that "dirty bomb" recipes can be found on al-Qaeda forums on the internet. This is slightly misleading, however. True, the manual *The Nuclear Bomb of Jihad* does mention the word "dirty bomb" (*qunbula qadhra*), but the kind of "dirty bomb" recipe it provides, consists of putting a piece of uranium "under the bed of the person you want to get rid of," claiming that this will kill him "instantly and without a scar."[24] This is of course utter nonsense. In the fall of 2006, however, a document entitled "Radioactive Pollution" was uploaded to the internet, which was of a very different category to the material discussed above. The two-page document starts with describing the effects of radioactive pollution, referring to two specific cases in which radioactive material was dislocated. The first incident took place in Goiania, Brazil in 1987, when a container of cesium-137 was stolen from an abandoned radium clinic and later dispersed. In Mississippi, USA the same year, small packages of thallium-67 and iodine-131 were spread along the highway due to a traffic accident. The description includes technical data on the substances involved, the extent of the pollution, as well as the costs of the clean-up. The aim of the author is, obviously, to show how even tiny amounts of such materials can cause major havoc and economical loss. He then provides suggestions on where such material can be "easily obtained," for example from "modern smoke detectors" and various medical equipment, noting that an easy and secure way of obtaining such material is "during its transportation between the place of production and the places of use or storage." Details on how to manufacture the RDD are not provided, the author simply suggests the user take the radioactive material and "put around it the explosives you have available." However, the method and rationale for such an attack is carefully explained:

> put the bomb in a city crowded with large markets and commercial shops. Explode it, even if it is time for the shops to close, in the evening for example, because the important thing is to spread the radioactive material in that commercial area, so that the government will close that area and everything around it because of the power of the material and the area of its dispersal. By this, you cause a large economic crisis to this country.[25]

Although such documents exist, it is remarkable how little interest is devoted to RDDs on the jihadi web, compared to the great attention given to this topic in Western media, as well as al-Qaeda's alleged ambitions in this direction.[26]

The nature of online discussions

Most of the CBRN manuals discussed above have existed for several years, and are circulated online as part of a larger jihadi "curriculum." Because of the large amount of material, however, it is hard to estimate the significance of these manuals. Are they actually being read or are they simply drowning in all the other material published? By looking at the nature of CBRN-related discussions taking place on jihadi forums, we might get further insight into al-Qaeda's CBRN intentions and capabilities. What are the main topics of concern for today's jihadists when it comes to CBRN weapons?

The first thing that can be noted is that forum members that engage in CBRN discussions do not appear to be highly trained or professional scientists. Rather, they seem to draw their knowledge from the media and the internet. In a discussion on ricin, for example, one of the participants stated that he had used Google to find information.[27] It is not surprising, therefore, that the most "popular" online recipes are those that are quick and easy and do not require advanced laboratory equipment, such as recipes inspired by *The Poisoner's Handbook*. When a member opened a discussion on biological weapons, encouraging others to ask him questions, another replied:

> Our *mujahid* brother ... are we able to manufacture them with simple equipment and available substances, and will they harm the person who

> manufactures them, and how far away are they able to harm our infidel enemies? ... Please teach us in a very simplified way, because we are not trained in biology.[28]

A commonly asked question is how to obtain the required chemicals or ingredients. Forum members sometimes enquire about the commercial name of chemicals, and what kind of stores to obtain them from. Due to a relatively high security awareness in these forums, the requests are seldom country- or region-specific, but there have been a few exceptions, with members stating they are from Palestinian areas or the Arab Peninsula.[29] Operational security in connection with obtaining chemicals has also been an issue of discussion. In one instance a forum member sought help to set up a laboratory without raising suspicion, saying:

> I have some theoretical knowledge of chemistry and I would like to develop it in practice, but buying the equipment requires me to sign papers, and I cannot ask someone else, because no one wants to get involved. Then, I will be discovered.

He received several suggestions on how to conceal his intentions, for example to start a laboratory for perfumes or chemical fertilizers, or to enroll in a scientific university.[30]

Safety precautions have also been a topic of discussion, especially when dealing with biological weapons. In a discussion started on *al-Firdaws* in November 2005, for example, several participants voiced concerns regarding the safety risks involved in the production of biological agents. One forum member regarded the subject as "very dangerous" due to the lack of isolation and containment procedures. Further he stated, "In case of a mistake or leak, it will become extremely dangerous for all Muslims in the area. It is not like an explosion, which is limited to the person working with it only."[31] The apparent unfamiliarity with biological agents, and the perception of them being deadly and uncontrollable, may also explain why the online interest in biological agents appears to be much smaller than for chemical ones.

One of the most common discussion topics is still the question on how to deliver or weaponize the agents. From July–November 2006, for example, there was a lengthy discussion on the *al-Firdaws* forum on how to manufacture and weaponize ricin. Significantly, one of the forum participants claimed to have already manufactured ricin and to have made experiments with its delivery. After a member wondered whether adding acid to ricin will produce ricin in gaseous form, he replied, "I poured concentrated sulphuric acid over the ricin poison but nothing happened, and no gases were produced, only the ricin turned brown."[32] Although displaying a rather shallow knowledge of chemistry, this is nevertheless a rare example of a delivery method actually being tested and reported on the forum. For the most part discussions on delivery methods remain theoretical, based on common knowledge, newspaper articles or other information found on the internet. Occasionally, forum members may also seek to discuss more non-traditional solutions such as delivering chemical weapons using remote-controlled model aircraft.[33] A more common suggestion is to weaponize the agent by simply mixing it with explosives, but no specific "recipes" for this seem to have been developed, and forum members have also expressed doubts with regards to the efficiency of this method. Interestingly, the alleged use of chlorine in Iraqi truck bombs during the first half of 2007 seems to be no topic of discussion, although it has received wide coverage in the media. One should keep in mind, however, that none of the jihadi groups present in Iraq have so far officially admitted the use of such bombs. As far as online jihadists are concerned, the alleged "chlorine bombs" are therefore viewed as part of a propaganda campaign to tarnish the reputation of the Islamic State of Iraq, rather than being a tactic currently employed by the Mujahideen.[34]

As mentioned above, nuclear weapons are quite a popular discussion topic. However, the lack of absolutely basic technical knowledge on the subject is striking. When asked how to protect oneself from radioactive material, for example, one "expert" answered that putting the radioactive material in the freezer will stop the radiation. Also, there are very few mentions of radiological weapons in the discussions. In general, internet users display poor knowledge on the nature of "dirty bombs," and usually suggest using a form of uranium in it, although natural uranium would be an ineffective material for this purpose due to its low radiation. Sometimes there are glimpses of realism in the discussions. One participant noted, for example, that a dirty bomb "will have a propaganda effect only."[35]

In at least a few cases the forum members have been encouraged to think in more realistic ways when it comes to selecting and delivering CBRN materials. In one instance, a member who proposed to spread ricin by placing the powder in the air conditioning vents of large buildings was given the following advice:

> My brother, don't get carried away ... my suggestion to you is to select a target, find its mailing address and write a threatening letter, then you put rat poison in the envelope and send it. A small step for you, but a large step for the Muslims.[36]

Although such examples are rare, there seems to be an increased awareness among jihadists that small-scale but feasible operations are better than large-scale operations with slim chances of succeeding. As one jihadist forum member recently put it, "a hand grenade that explodes in one of New York's streets, is better than a nuclear bomb capable of destroying half of New York that does not explode!"[37]

Conclusion

In general, this analysis has revealed two important findings: first, that the subject of CBRN weapons receives very limited attention among al-Qaeda's followers online; and second, that there is an apparent lack of knowledge and innovative ability regarding CBRN-related means and methods on the part of the jihadists. If the online training and instruction material on jihadi websites is an indicator of the state of al-Qaeda's CBRN capability, the terrorist network will at most be able to deploy a *crude chemical device*, which will have few similarities with militarily effective chemical, biological or nuclear weapons.

True, if al-Qaeda were in possession of a more advanced CBRN capability, they would hardly divulge this on jihadi internet forums, which after all are known to be monitored by various Western and Arab intelligence services. It is also possible that there are more sophisticated CBRN manuals and discussions in other and less accessible parts of the jihadi web that admittedly lie beyond the reach of most academic researchers. The threat of a CBRN terrorist attack from al-Qaeda can therefore not be dismissed based on this study only.

What do these findings mean then, in terms of future threat assessments? Clearly, the manuals and discussions by themselves can hardly be used to identify specific threats. For this purpose the CBRN recipes are too crude and ineffective, and the discussion forums are in any case not used to exchange sensitive information. What is interesting, however, is that this online literature seems to confirm a trend which is already evident on the ground, namely the preference for crude CBRN materials and low-tech delivery methods. After 2001 jihadi cells in Europe reportedly planned to carry out low-end chemical and radiological attacks, relying on materials such as castor beans and commercial smoke detectors, and in 2007 Iraqi insurgents started to combine truck bombs with chlorine canisters. The crude nature of online manuals and discussions seems to confirm this picture.

When assessing the future CBRN threat from al-Qaeda, therefore, we should not simply focus on the most dangerous CBRN weapons of top military quality and effectiveness, although these definitely represent a grave threat. It is equally important to be aware of the developments actually taking place today and to consider how these might develop. This points to a future of medium- and low-level threats, emanating from simple but inventive use of various chemical, biological or radiological materials. Even though Shaykh al-Fahd's fatwa from 2003 has attributed some sort of "religious legitimacy" to the use of WMD to kill millions in one stroke, this seems to have received much more attention among Western policy-makers than among al-Qaeda's own followers. By focusing blindly on "high-tech" terrorism we might underestimate terrorists' innovative capabilities and, consequently, fail to recognize other types of scenarios that, after all, appear far more likely.

Notes

1 This chapter is an extended and updated version of my article in *Jane's Intelligence Review*, September 2007, entitled "Chem-bio Cyber-class: Assessing Jihadist Chemical and Biological Manuals."

2 Nasir bin Hamd al-Fahd, "A Treatise on the Legal Status of using Weapons of Mass Destruction Against Infidels," (in Arabic), Rabi'i 1423h (May 2003). See translation and

analysis in Reuven Paz, "YES to WMD: The First Islamist Fatwah on the use of Weapons of Mass Destruction," *PRISM Series of Special Dispatches on Global Jihad*, 1. Online at: www.e-prism.org/images/PRISM%20Special%20dispatch%20no%201.doc (accessed October 2004).

3 Some exceptions are: Sammy Salama and Lydia Hansell, "Does Intent Equal Capability? Al-Qaeda and Weapons of Mass Destruction," *Nonproliferation Review*, 12:3 (2005), pp. 625–6; Adam Dolnik and Rohan Gunaratna, "Jemaah Islamiyah and the Threat of Chemical and Biological Terrorism," in R. Howard and J. Forest (eds.), *Weapons of Mass Destruction and Terrorism* (USA: McGraw-Hill, 2008).

4 FFI has undertaken several studies on al-Qaeda and the internet. See, for example: Hanna Rogan, "Jihadism Online – A Study of How al-Qaeda and Radical Islamist Groups use the Internet for Terrorist Purposes," FFI report-2006/00915. Online at: http://rapporter.ffi.no/rapporter/2006/00915.pdf (accessed March 30, 2007); Brynjar Lia, "Al-Qaeda Online: Understanding Jihadist Internet Infrastructure," *Jane's Intelligence Review*, December 2, 2005. Online at: www4.janes.com/subscribe/jir/doc_view.jsp?K2DocKey=/content1/janesdata/mags/jir/history/jir2006/jir01397.htm@current&Prod_Name=JIR&QueryText= (accessed March 30, 2007).

5 See for example: Stephen Ulph, "A Guide to Jihad on the Web," *Terrorism Focus*, 2:7. Online at: www.jamestown.org/terrorism/news/article.php?articleid=2369531 (accessed February 28, 2007).

6 In three other forums examined, the percentage of threads dedicated to training and preparation were as follows: *al-Nusra*: 2.8 percent; *Abu al-Bokhary*: 3.9 percent; *Risalat al-Ummah*: 3.9 percent (accessed December 19, 2006).

7 Homepage of the *Encyclopaedia of Preparation*: www.geocities.com/m_eddad/ (3rd edn); and www.geocities.com/i3dad_jihad4/ (4th edn) (accessed March 30, 2007). The links on the page did not work at the time of access. The *Encyclopaedia of Preparation* should not be confused with the *Encyclopaedia of the Afghan Jihad* (a nine- or 11-volume collection of training manuals issued by the Office of Services in the early 1990s) or the *Encyclopaedia of Jihad* (a two-volume electronic version of part of the *Encyclopaeda of the Afghan Jihad*). These documents are, however, largely included in the *Encyclopedia of Preparation* collection.

8 The video, entitled *The Poor Man's James Bond Greets the Russians*, was made by the American "survivalist" Kurt Saxon (born Donald E. Sisco) and can be purchased from Saxon's homepage, www.kurtsaxon.com (accessed March 30, 2007). The video is based on a recipe in *The Poor Man's James Bond*, Vol. 3, also written by Saxon and available from Desert Publications.

9 For more information on al-Qaeda's online training manuals and videos, see: Anne Stenersen, "The Internet: A Virtual Training Camp?" *Terrorism and Political Violence* (2008, forthcoming).

10 The forum *al-Firdaws* ("The Paradise") was selected for this analysis because it had the largest archive of threads in addition to having a separate sub-forum for equipment and preparation. The archive dates back to February 2005.

11 Maxwell Hutchkinson, *The Poisoner's Handbook* (El Dorado: Desert Publications, 2000). The book was first published in the 1980s.

12 *The Mujahideen Poison's Handbook*, p. 1.

13 Maxwell Hutchkinson, *The Poisoner's Handbook*; Kurt Saxon, *The Poor Man's James Bond Greets the Russians*.

14 René Pita, Juan Domingo Álvarez, Carmen Aizpurua Sánchez, Sergio González Domínguez, Alberto Cique Moya, José Luis Sopesen Veramendi, Matilde Gil García, María del Valle Jiménez Pérez, Carmen Ybarra de Villavicencio, Juan Carlos Cabria Ramos, Arturo Anadón Navarro, "Extraction of Ricin by Procedures Featured on Paramilitary Publications and Manuals Related to the Al Qaeda Terrorist Network," (in Spanish) *Medicina Militar*, 60:3 (2004).

15 Maxwell Hutchkinson, *The Poisoner's Handbook*, pp. 25–6.

16 *The Poisons Encyclopaedia*, p. 14 (PDF version). "The above-mentioned symptoms" refer to nausea, vomiting and paralyzation, then blurred vision, dilated eye pupils, relaxed muscles, severe headache, deepening and then disappearance of the voice, and usually high temperature.
18 "Preparation of Botulism TOXIN," unknown author. Posted by "bio man" on Yahoo! Groups, January 8, 2005.
18 Names of some of the chemicals have been omitted.
19 *The Unique Invention*, p. 1.
20 This author's version of *The Unique Invention* was downloaded by the Norwegian Defence Research Est. (FFI) in the fall of 2004. The file was created on November 3, 2003. Ron Suskind claimed in 2006 that an al-Qaeda cell had planned to use the device against the New York Metro in 2003, but that the plan had been aborted by Ayman al-Zawahiri. The claim is hard to verify. The information was given by a CIA informant and has not been confirmed by other open sources. A manual describing the device was also found on a computer belonging to an al-Qaeda suspect arrested in Bahrain, January 2003, but there were no indications that the arrested individual had made experiments with the device. Ron Suskind, *The One Percent Doctrine* (London: Simon & Schuster, 2006).
21 "On Islamic Websites: A Guide for Preparing Nuclear Weapons," MEMRI Special Dispatch Series – No. 1004, October 12, 2005. Online at: http://memri.org/bin/articles.cgi?Page=archives&Area=sd&ID=SP100405 (accessed July 9, 2007). The article describes a manual with the title *The Nuclear Bomb of Jihad and How to Enrich Uranium* and the summary of the content appears to correspond to the content in the manual reviewed by this author.
22 *The Nuclear Bomb of Jihad and the Way to Enrich Uranium*, p. 2.
23 The concept could be carried out in practice, but would require large industrial complexes. For example, 13,000 tons of sand would have to be processed to obtain enough uranium for one nuclear weapon. See "Production of Nuclear Materials," *Nuclear Threat Initiative*. Online at: www.nti.org/e_research/cnwm/overview/technical4.asp (accessed July 9, 2007).
24 *The Nuclear Bomb of Jihad*, chapter 18.
25 "Radiological Pollution," pp. 1–2.
26 See for example, E. Lederer, "UN Paints Grim Portrait of Qaida Threat," *Associated Press*, November 28, 2007. Online at: http://ap.google.com/article/ALeqM5hKcDTXO9LLYXtIUQd6ag9zPdAGNAD8T6A7QO0 (accessed November 30 2007).
27 See for example, "Ricin Seeds ... Praise God," (in Arabic) thread started by "hamoda" on *al-Firdaws*, July 9, 2006. Online at: www.alfirdaws.org/vb (accessed January 10, 2007). One forum member says, "I advise you, my brother, to search in Google to learn about this substance and its importance and risk." Another member later reports, "[T]his is what I found from Google."
28 "Course on biological weapons," (in Arabic) thread started by "modjahede" on *al-Firdaws*, November 8, 2005. Online at: www.alfirdaws.org/vb (accessed January 10, 2007).
29 See for example, "Chemical weapons," (in Arabic) thread started by "azzam 2000" on *al-Firdaws*, July 5, 2006. Online at: www.alfirdaws.org/vb (accessed September 22, 2006).
30 "How do I set up a laboratory without raising suspicion? Asking help," (in Arabic) thread started by "sajy_79" on *al-Nusra*, October 25, 2006. Online at: www.alnusra.net/vb (accessed November 19, 2006).
31 "Course on biological weapons," (in Arabic) thread started by "modjahede."
32 "Ricin Seeds ... Praise God," (in Arabic) thread started by "hamoda."
33 "Important question," (in Arabic) thread started by "saifo allah" on *al-Firdaws*, June 8, 2005. Online at: www.alfirdaws.org/vb (accessed January 10, 2007).

34 See for example, "'Marathon' of lies about the Islamic State of Iraq," (in Arabic) thread started by "mujahidat al-sham" on *al-Ekhlaas*, May 15 2007. Online at: www.alekhlaas.net/forum/showthread.php?t=59489 (accessed July 12, 2007). The Islamic State of Iraq is the name of an umbrella organization for several insurgent groups, including Iraq's local branch of al-Qaeda.

35 "layth al-islam," "The Nuclear Bomb of Jihad," *al-Firdaws*, October 6, 2005. Online at: www.alfirdaws.org/vb/showthread.php?p=16731&page=2 (accessed May 22, 2006). However, the same user displays a lack of knowledge about the nature and characteristics of RDDs. The message started with the following definition:

> The dirty bomb is a bomb made of regular explosives, where radioactive material has been added, like we add to the thermit uranium oxide with iron oxide and pieces of aluminium, knowing that radioactive materials consist of all the elements following after mercury in the periodic table.

36 "Ricin Seeds ... Praise God," (in Arabic) thread started by "hamoda."

37 "Urgent message to the blessed jihadi cells," (in Arabic) thread started by "ibn al-tanzim" on *al-Ekhlas*, July 1, 2007. Online at: www.alekhlaas.net/forum (accessed July 2, 2007).

Part III

CBRN, capacity-building and proliferation

4 Indicators of chemical terrorism

Amy E. Smithson

Introduction

Though circumstances can propel terrorists to innovate, they have proven to be strongly reliant on familiar tools to inflict harm and induce fear. For a while, kidnappings, airline hijackings and armed attacks were often terrorists' tools of choice, but in the past decade terrorists have turned most often to the use of bombs. Whether placed in vehicles (e.g. trucks, boats), planted by the roadside and triggered remotely or strapped to individuals intent on sacrificing their lives, the statistics about bombs tell the tale. The terrorist use of bombs was so preponderant from 1997 to 2006 that bombings comprised 12,806 out of a total of 23,135 attacks, just over 55 percent of all attacks. Moreover, bombings were more than twice as prevalent as the next most frequently employed terrorist method of attack, the use of guns.[1]

Following Aum Shinrikyo's mid-March 1995 attack with the nerve agent sarin on the Tokyo subway, many security analysts heralded a new era of catastrophic terrorism where terrorists would increasingly employ unconventional weapons such as chemical and biological agents and radiological dispersal devices.[2] Allotting terrorists two years to make the transition to widespread use of unconventional weapons, terrorists still executed relatively few unconventional attacks from 1997 to 2006. Twelve of the 19 biological events in this period were a string of attacks in 2001 that killed five people through the mailing of anthrax-laced letters to US media organizations and elected officials. The remaining biological events, so-called "white powder incidents," caused disruption but no harm. No radiological attacks took place during this time period. The 27 chemical events that occurred from 1997 to 2006 involved the dispersal of non-lethal substances (e.g. tear gas), the throwing of acid on individuals and the sending of letters containing traces of ricin and cyanide. Combined, these chemical attacks caused a few injuries and deaths. The most harmful event was a food-poisoning incident, when cyanide put into a curry dish at a 1998 street festival in Wakayama in Japan killed six and sickened 60.[3]

The statistics related to terrorist use of unconventional weapons since 1995 indicate that those forecasting catastrophic terrorism were more preoccupied with these weapons than were the terrorists themselves. Specifically with regard to chemical weapons, these statistics beg important questions, including the factors possibly holding terrorists back from turning to chemical weaponry and those that might propel terrorists to more frequent use of poison gas. The balance of this chapter first considers the characteristics of chemical weapons that might induce terrorists to incorporate them into their regular attack toolkit. The chapter then dissects the steps involved in the terrorist acquisition and utilization of chemical weapons, identifying along the way the visible indicators of such activity. Two case studies, Aum Shinrikyo and al-Qaeda, are used for this purpose. Various factors that might play a role in the terrorist acquisition of chemical arms are then considered, including the possible interaction of terrorist groups with state sponsors of terrorism and technical advances that could bring this category of weaponry more within the reach of sub-national actors. Two such advances, namely novel warfare agents and microreactors, are explored. The chapter concludes with a discussion of a shortcut that terrorists might take to create a chemical disaster and some concluding thoughts about indicators of terrorism and a string of attacks in 2007 that could denote that terrorists are reconsidering the more common use of chemicals for attacks.

The lure of chemical weapons as a terrorist tool

Death by any means is an unpleasant thing to contemplate, but one of the more horrific methods of perishing has to be from something that can be invisible and odorless yet that suddenly leaves victims with searing skin and lungs, gasping for air and convulsing. A nerve agent can cause death within minutes. Even if a chemical agent is dispersed in a concentration dense enough to cause a visible cloud or has an odor (e.g. a garlic-like smell), victims will realize that something in the air is poisoning them.[4] Few things are more terrifying than a poison gas attack; bullets and bomb fragments might be dodged, but humans know they have to breathe.

Information about the effects and effectiveness of choking, blood, blister and nerve agents has long been in the public domain. Like conventional bombs, chemical warfare agents harm quickly, which is one of the reasons terrorists may find them attractive. Moreover, terrorists can tailor an attack by selecting a chemical agent that kills within a couple of minutes (e.g. VX) or one that acts more slowly, is not as lethal and may instead leave victims with debilitating health conditions (e.g. mustard gas) if dispersed in a moderate concentration. Accordingly, terrorists can choose to kill plenty or to slay some, injure masses and, via the subsequent media coverage, frighten copious others. Executing a chemical attack in an ordinary setting such as a transportation system, a shopping mall or a school can heighten the sense of vulnerability that could engulf the public after such an attack. Part of the appeal of chemical agents is that their variety and their potential dispersal flexibility could allow terrorists to dial up or down the extent of damage an attack inflicts.

Another part of the allure that chemical weapons might hold for terrorists rests in the relative ease and low cost of acquiring knowledge, materials and equipment relating to this category of weaponry, particularly in comparison to nuclear weapons. Beginning in the early 1900s, multiple variants of the formulas for chemical warfare agents were published in the patent literature and textbooks. More recently, "cookbooks" such as Uncle Fester's *Silent Death* have made technical information available. Some of the data in these books is sound, some of it is not.[5] Expert opinion on the level of education or training needed to make chemical warfare agents differs; someone with high-school training in chemistry can probably make modest quantities of an agent, but a person with a master's degree in chemistry and/or industrial experience working with highly corrosive, hazardous chemicals is likely run to a safer operation that yields a better product and larger quantities of the agent.[6]

Basic equipment such as reactors, control units, piping and storage vessels can be purchased new or used on the open market. State-of-the-art equipment such as glass- or alloy-lined reactors and pipes that are more resistant to corrosion can be employed, but agents can also be made with less expensive options, including beakers and plastic tubing for smaller quantities. Some supplies (e.g. muriatic acid, sulfuric acid) are available at pool supply companies and even in products on the shelves at grocery stores. In small amounts, some direct precursor chemicals can be bought from chemical suppliers without drawing much attention. Or, terrorists could buy larger quantities of building-block chemicals and produce direct precursors and ultimately warfare agents in-house. Estimates for setting up a full-scale chemical agent manufacturing plant range from $5 million to $20 million,[7] but smaller amounts of agent can be made with a more modest investment in infrastructure. If purchases are handled cleverly and the manufacturing plant is situated amidst the clutter of an industrial area or in a remote location, a terrorist group could set up and operate a small-scale chemical agent production shop without drawing much suspicion.

To disperse chemical agents terrorists can, of course, employ munitions (e.g. bombs, rockets, artillery). If military weapons designed to deliver chemical agents are not accessible, terrorists could fashion homemade explosive dispersal systems, mindful of the appropriate explosive–agent ratio necessary to disseminate rather than destroy the agent. Other delivery options include attaching sprayers to aircraft or vehicles. Sprayers can also be directed into the ventilation systems of buildings or positioned in other key locations at a site. Farm equipment stores sell large sprayers; small ones (e.g. backpack or reusable canister systems) can be found in ordinary hardware stores. In short, the available delivery options range from dedicated military systems to low-tech and homemade.

When accessibility of equipment, materials and know-how is taken into consideration, chemical agents appear to be the low-hanging fruit of unconventional arms. Perhaps only the assembly of a radiological dispersal device could be considered easier than creating a rudimentary chemical weapons capability. For terrorists, use of poison gas opens up options ranging from thousands of casualties to the mass terror of a well-placed and well-executed small-scale attack. Like bombs, the quick-acting chemical agents create an instant impact and dramatic pictures, just the type of result publicity-hungry terrorist groups seek. Moreover, the physical and psychological after-effects of unconventional weapon attacks can be much more pronounced than those from conventional terrorist attacks.

Case studies in chemical terrorism

To those who suffer the brunt of attacks, terrorist strikes appear to come out of nowhere, but that is hardly the case. The element of surprise is indeed an integral part of terrorism, particularly the exact timing and targeting of attacks, but terrorism is a familiar foe that dates to Biblical times. As such, terrorism has identifiable characteristics, from the spawning and spreading of the ideas that generate attacks to the series of activities that turn those ideas into awful reality. Some of those activities will be visible to the meticulous observer, others probably not unless law enforcement or intelligence authorities infiltrate terrorist groups.

The following section of the chapter traces the chemical weapons-related activities of two terrorist groups, Aum Shinrikyo and al-Qaeda, visible to outside observers. Such signs can be picked up by analyzing the publicly released materials of these groups and closely watching the activities of its members. Detectable signals can be found in all manner of ways – for instance by pursuing a group's purchasing activities or by keeping a sharp eye out for individuals loitering in subways rather than commuting. As always, the challenge for intelligence and law enforcement agencies is to collect and separate the meaningful data from the background noise of other activity that is of no consequence to efforts to identify chemical terrorists in the making and prevent their plans from materializing.

Indicators of Aum Shinrikyo's chemical weapons activities

Aum Shinrikyo's attack on the Tokyo subway on March 20, 1995 stunned practically everyone except the perpetrators of the attack and the Japanese law enforcement authorities who had gradually begun to suspect the cult of various nefarious activities, including the manufacture and use of poison gas. The cult's religious status gave it legal protections that required Japanese police to investigate with extra caution. When Japanese law officials finally had sufficient evidence to clamp down on the cult due to suspicion of kidnapping, an insider tip to Aum Shinrikyo about a planned March 22, 1995 police raid on the cult's compound prompted Aum to lash out against the police. The subway lines that Aum selected for its infamous sarin attack intersected at Kasumigaseki, the headquarters location of Japan's law enforcement authorities, and the attack was timed for when shifts would be changing in order harm the maximum number of police.[8] Ordinary citizens, not the police, fell victim to the sarin, with a dozen killed, 54 seriously or critically wounded, 984 mildly injured and countless other commuters, by-standers and members of the general public terrorized.[9]

The cult's leader, Chizuo Matsumoto, was the mastermind, though not the technical muscle, behind Aum Shinrikyo's foray into unconventional terrorism.

A partially blind, charismatic individual, Matsumoto renamed himself Shoko Asahara or "Bright Light." Asahara's first gambit to expand Aum's influence was a miserable embarrassment when the cult's slate of 25 Truth Party candidates did not win election to the Japanese parliament. Thereafter, Asahara plotted to subvert the Japanese government. To convince his followers that they needed to arm themselves for cataclysmic battle, Asahara created a caustic mix of the revelations of Nostradamus, the Christian concept of Armageddon and the Buddhist notion of anarchy following the abandonment of Buddha's wisdom. Asahara published his apocalyptic philosophy, including twin volumes in 1989, *The Day of Destruction* and *From Destruction to Emptiness*, and the 1995 book *Disaster is Approaching in the Country of the Rising Sun*, which describes Asahara's views on biological weapons.[10] In 1994 the cult also released a video entitled *Slaughtered Lambs* to buttress its claims that US and Japanese aircraft had attacked its compounds with poisonous substances. Moreover, Asahara publicly asserted that the poison gas attack on Matsumoto, described briefly below, was really the work of US Navy pilots.[11]

In addition to chemical warfare agents, Aum hoped to produce lasers and biological weapons. Asahara's senior lieutenants publicly glorified Ebola as a weapon.[12] Materials in the public domain – statements by cult officials, Asahara's books, the video – were there in abundance to make law enforcement and intelligence officials aware of the cult's ominous guiding philosophies, its plans and its activities.

While Aum's biological weapons program was a start-to-finish failure,[13] Aum's chemical weapons program was much more successful and ended up making the cult notorious. Before creating an in-house manufacturing capability, Aum Shinrikyo tried to buy chemical weapons off-the-shelf from known possessors. In 1988, a South Korean operative approached a man in a bar in Utah asking about buying US chemical bombs. This individual tipped off US Customs and Border Protection, which subsequently contacted the Aum operative posing as a company that sold American weapons. The cult placed an order for 500 each of MK94 bombs and MK116 "weteye" bombs. These munitions contained 250 tons of sarin.[14] The South Korean operative apparently believed he could load these bombs into the trunk of his car, not realizing that their size is such that four weteye bombs in their protective packing fill a railroad boxcar.[15] US Customs officials foiled this clumsy procurement attempt, but apparently did not share news of it with US law enforcement and intelligence officials or those officials did not deem the incident sufficiently alarming to look further into the cult's activities. US intelligence and law enforcement agencies testified that prior to March 20, 1995 US officials did not view the Japanese cult with undue suspicion or as a danger, despite the fact that a January 1995 edition of the cult's propaganda newsletter alluded to a threat on the life of US President Bill Clinton, and the fact that the cult repeatedly portrayed the United States as it's enemy in written and oral statements.[16]

Although Aum's bumbling effort to procure US chemical weapons fell short, the cult also approached Russia's State Scientific Research Institute of Organic

Chemistry and Technology in the hopes of buying munitions or gaining chemical weaponeers as recruits. Aum did not come away with any Russian chemical rounds or weaponeers. However, the cult is reported to have paid Oleg Lobov, the Secretary of the Russian Security Council, $100,000 for the blueprints that guided the construction of its main sarin production plant, which was located at their Kamikuishiki compound at the base of Mt. Fuji.[17] Whether from the Russians or some other supplier, Aum also managed to obtain a Russian-made GSP-11 toxic gas detector.[18]

Unable to buy ready-made weapons, Aum developed its own chemical warfare program. The cult spent approximately $30 million between 1985 and 1995 to develop, test and produce chemical agents.[19] Over $10 million of this amount went to Satyan 7, the sarin production plant, which included three laboratories, a computer control center and five reactors, all top-flight Hastelloy alloy models, one of which was a $200,000 Swiss-made computerized reactor with automatic temperature control and record-keeping capabilities.[20] The cult's production goals were consistent with an intention to cause mass casualties. Aum aimed to produce 70 tons of sarin in 40 days, creating two tons of sarin per day in batches of 17.6 pints.[21]

Once the cult embarked on the effort to devise various weapons itself, Asahara gave a standing order to recruit top scientists and technicians to its Clear Stream Temple, the cult's high-technology weapons division at its Mt. Fuji compound. In addition to universities and other high-technology workplaces, Aum Shinrikyo also considered the Japanese Self Defense Forces as a recruitment ground. Among those lured to the cult was Masami Tsuchiya, a former PhD candidate at Tsukuba University who became the lead chemist for Aum's program. When the cult stepped up the pace of operations at Satyan 7, the number of scientists, engineers and technicians working in the chemical weapons program exceeded 100.[22]

While it botched attempts to buy weapons, Aum proved amply capable of acquiring the building-block and precursor chemicals to feed the fabrication of chemical warfare agents. The cult established two front companies, Bell Epoch and Hasegawa Chemicals, through which it purchased over 200 different chemicals from more than 200 Japanese chemical supply companies. Among other chemicals, two days after the subway attack Japanese law enforcement authorities seized from Aum's warehouses:

- 60 tons of glycerol;
- 50 tons of phosphorous trichloride;
- 1.5 tons of nitric acid;
- 1.2 tons of calcium chloride; and
- smaller quantities of dimethylamino ethanol, dimethylmethylphosphonate, hydrogen fluoride, methyliodide, potassium iodide, thiodiglycol and sodium cyanide.[23]

In addition to the cult's apocalyptic philosophy and the quantity and type of the chemical purchases, Japanese police might have been tipped to the cult's malevolent intentions by the fact that for several years neither Aum Shinrikyo's front companies nor its other enterprises made any products with these chemicals. With the significant array of equipment and chemicals that Aum Shinrikyo bought, the lack of any viable commercial output should have led someone to question the purpose of the group's investment if not for commercial products. Such questions would have been all the more obvious to ask had law enforcement or intelligence authorities picked up on the cult's purchases of chemical antidotes and personal protective equipment.[24]

Aum's scientists conducted laboratory toxicity tests of their biological agents using small rodents and probably did the same to assure the lethality of their chemical warfare agents.[25] Absent an accident that caused a leak of toxic fumes from the laboratory or caused significant casualties among workers, there would be few visible indications of testing activities inside of a laboratory for law enforcement or intelligence officials to detect. However, to test the toxicity of its agents, in September 1993 Aum purchased a ranch in Australia called Banjawarn Station. In outdoor toxicity tests, the cult killed at least 29 sheep with the nerve agent sarin. Proof positive of this activity came when Australian authorities found residues of sarin on the remains of the sheep and in the soil at the ranch. After the Tokyo subway attack, Australian officials also came across Japanese-language documents about the field toxicology tests.[26] In the spring of 1994, Aum Shinrikyo's live-agent testing continued with repeated assassination attempts against its rivals. For example, Aum used sarin to attack Daisaku Ikeda, head of the Soka Gakkai Buddhist organization. Attempting to spray sarin from a device affixed to a van, in one attack the device sprayed backwards.[27] Aum thrice used the nerve agent VX in assassination attempts against other opponents, killing one.[28]

On June 27, 1994 the cult upped the ante of its chemical warfare attacks. Asahara ordered an assault on three district court judges in Matsumoto to prevent an anticipated unfavorable ruling in a civil suit about the cult's purchase of a piece of land.[29] The attack almost went dangerously wrong for the cult members, who used too much isopropyl alcohol with the sarin drip, causing their van to fill with a white mist of sarin. With respiratory protection consisting of inflated plastic bags with oxygen feed tubes, the cult members were fortunate to have taken the sarin antidote prior to the attack. The cult members sprayed for 20 minutes and dispersed about 20 kg of sarin.[30] The Matsumoto attack did not harm the judges, but it killed seven people, hospitalized 51 and sent 253 to outpatient clinics.[31] At this point, chemical warfare agents had been used to slay eight Japanese citizens and seriously injure dozens of others. The unusual weaponry in these assaults was clearly identifiable.

Though Japanese police and media initially pinned responsibility for the attack on a resident of the gassed Matsumoto neighborhood, before long leads began pointing to Aum Shinrikyo. In September 1994, a major Japanese media organization received an 11-page statement asserting that Aum perpetrated the attack. A second anonymous letter claimed that Matsumoto was an open air "experiment of sorts" and that the results would be much worse if the sarin was released indoors, for example, in a "crowded subway."[32] Meanwhile, as the cult began trying to increase the output of Satyan 7, citizens living near the Kamikuishiki compound complained of noxious fumes. In November 1994, police sampled the soil around Aum's compound and analysis detected traces of chemicals in proportions consistent with the final chemical reaction to manufacture sarin.[33] The field toxicity tests, the assassination attempts, and the soil sample analysis were glaring indicators that the cult had successfully made chemical warfare agents and was willing to use them on animals and people alike.

Other noticeable indicators of Aum's malevolent intentions were the cult's several attempts to spray its biological materials. In mid-1994, Aum twice tried to spray an anthrax concoction from the roof of its Tokyo headquarters building. A police report of the first incident notes that those living near the building smelled something malodorous, and according to widely repeated accounts, cult members wearing moon suits were seen pouring a solution into a pump and then through the sprayer and fan.[34] On other occasions, Aum sent a vented van with a compressor pump around Tokyo to spray supposed botulinum toxin and anthrax solutions and reportedly tried to spray the Japanese Diet with a neurotoxin using a homemade device.[35] On March 15, 1995, just before the subway attack, Aum placed three briefcases with side vents, battery-powered fans and vinyl tubes to contain biological agent in the Kasumigaseki Station of the Tokyo subway. The vials in the briefcases, which authorities found, were empty.[36] Spraying of foul-smelling substances and leaving homemade dispersal devices in the very subway station that was the focal point of the March 20, 1995 attack were suspicious activities visible to law enforcement and intelligence officials.

Once Aum was tipped that the police were about to raid its compound, Asahara directed cult members to lay the groundwork to preempt the police with the subway sarin attack. Though planned quickly, Aum's sarin attack plan was methodical. On March 19, cult members could be seen scouting the Marunouchi, Hibiya and Chiyoda lines, which crossed at the Kasumigaseki Station. The cult members were recording the frequency of the subway trains, their arrivals and departures, the length of time that the train doors remained open and the positioning of subway cars closest to the exits from the stations. Late that night, Aum attempted to distract the police by firebombing its own headquarters and dropping leaflets blaming the attack on a rival cult.[37] The next day, after numerous visible activities acquiring and using chemical warfare agents, Aum Shinrikyo acted in concert with Asahara's apocalyptic philosophy and prophecies. Following the plan that gelled after the subway reconnaissance mission, between 7:46 a.m. and 8:01 a.m., Aum's designated attackers boarded five trains, punctured sarin-filled packages with the tips of umbrellas, and exited the trains several stops before the Kasumigaseki Station. With that attack, Aum created its own despicable chapter in the annals of terrorism.

Indicators of al-Qaeda's activities

Born amidst the Muslim opposition to the 1979 Soviet invasion of Afghanistan, the terrorist network al-Qaeda refocused its fury when the Coalition military presence in the Middle East during the 1991 and 2003 Gulf Wars generated fuel for hostility toward the West. Al-Qaeda, Arabic for "The Base," seeks to supplant Western cultural and political influences with fundamentalist Islamic practices and governance. Accordingly, al-Qaeda's political and religious leaders have declared a "jihad," or struggle against the infidels of the West.[38] Moreover, al-Qaeda's political and religious leaders have issued "fatwas," or rulings that have the effect of edicts within the Muslim community, that buttress the declaration of jihad. Al-Qaeda's leader, Usama bin Laden, issued fatwas in 1996 and 1998 authorizing the killing of American soldiers and civilians, as well as those of allied countries, effective until the withdrawal of Western forces from Muslim holy lands is complete and US and allied support for Israel ceases.[39]

Al-Qaeda and its sister organizations have splintered into over 50 countries, but the size of the organization is difficult to estimate because of the independence with which jihadi cells operate. To communicate effectively with its followers and potential recruits, al-Qaeda's leadership has taken increasing advantage of modern multimedia technologies (e.g. videotaped messages sent to television networks such as al-Jazeera, use of the internet), which means that its philosophies and directives have been readily available to the law enforcement and intelligence communities. Consistent with the jihad and fatwa declarations, al-Qaeda has gravitated toward attacks on high-profile Western targets of opportunity (e.g. military assets, landmarks, transportation nodes) that have sometimes caused widespread and indiscriminate loss of life. In addition to the attacks of September 11, 2001, al-Qaeda has claimed responsibility for or has been accredited with the August 7, 1998 simultaneous bombing of US embassies in Kenya and Tanzania, the March 11, 2004 bombing of trains in Madrid and the July 7, 2005 bombing of the London subway, among other attacks.

More specifically with regard to unconventional weapons, in December 1998 interviews with Western media outlets, bin Laden literally instructed his followers to acquire weapons of mass destruction, characterizing that activity as a "religious duty" for Muslims.[40] Another noteworthy fatwa, issued by Sheik Nasir bin Hamd al-Fahd in May 2003, provided Muslims permission and a rationale to use weapons of mass destruction against Americans.[41] Bin Laden bragged in November 2001 that al-Qaeda had succeeded in acquiring both chemical and nuclear weapons "as a deterrent" against America.[42]

For years, al-Qaeda members received hands-on instruction in terrorist tactics in boot camps located mostly in Afghanistan, but more recently recruits willing to travel to Iraq have had ample opportunity to learn, refine and employ fighting skills in the ongoing insurgent campaign in Iraq. Followers can also take instruction by consulting al-Qaeda's massive *Encyclopedia of Preparation for Jihad*, some of which is available online. The eleventh volume of this encyclopedia focuses on how to make chemical and biological agents and includes a formula for making sarin. Though the entire encyclopedia is large in size, the quantity of the unconventional weapons instruction in this manual is not that large and its quality is of questionable utility. Moreover, the unconventional weapons material appears to be of much less interest to aspiring jihadists than the encyclopedia's tutorials on conventional attack tactics.[43]

In addition to these visible indicators of al-Qaeda's interest in the acquisition and use of chemical weapons, evidence has come to light that shows actions to translate that interest into reality. After the fall 2001 invasion of Afghanistan by US and Coalition troops, US troops found traces of precursor chemicals for a blister agent at one al-Qaeda compound, some chemical equipment and materials related to poison gas weapons in Afghanistan.[44] Plans and a record of activities related to al-Qaeda's chemical and biological weapons program were also recovered in documents and on a computer. This weapons program, codenamed "Curdled Milk," reportedly involved plans and materials to make cyanide as well as efforts to manufacture a nerve agent or pesticide that was reportedly tested on mid-sized animals (e.g. rabbits, dogs).[45] These tests may have been those seen in a video that the television network CNN released in August 2002. In the video, three dogs succumbed to the effects of a toxic substance, maybe a chemical warfare agent but perhaps an industrial organophosphorous chemical.[46] The US assessment in the fall of 2002 was that the captured information and materials indicated that al-Qaeda was likely to have had a limited capacity to make chemical agents.[47]

Aside from these activities, additional indicators of al-Qaeda's efforts to wield chemicals as a weapon can be found in plots that law enforcement authorities have foiled. British officials interrupted a plot in April 2004 to marry osmium tetroxide with conventional explosives, creating a chemical dirty bomb apparently targeted for the transportation system and shopping centers in London.[48] That same month in Amman, Jordanian officials prevented an apparent plan to blow up two chemical-laden vehicles at the Jordanian General Intelligence Department. Initial reports claimed that the attack would have killed thousands and that authorities found 20 tons of assorted chemicals in the possession of al-Qaeda sympathizers. The evaluation of outside analysts, however, was that the attack would have produced a toxic mess but probably not mass casualties from poisoning.[49]

The jihad declaration and fatwas, as well as the accessibility of applicable materials and knowledge, will persist indefinitely, making it likely that al-Qaeda faithful will continue to pursue chemical weapons. Principal al-Qaeda talent that enabled the group's more notorious attacks can no longer contribute to such efforts. Abu Khabab al-Masri, one of al-Qaeda's top explosives experts who trained recruits in Afghanistan in the manufacture and use of poisons and also wrote many of the terrorist group's instructional materials related to chemical and biological weapons, was killed in early 2006.[50] Other al-Qaeda top brass no longer available to serve the cause include the plotter of the September 11 attacks, Khalid Sheik Mohammed, and field commander Abu Zubaida.[51] Though the ranks of al-Qaeda's senior veterans have thinned, the organization has enrolled middle-class, educated individuals with technical, linguistic and cultural skills that could make them more likely to blend into Western society.[52]

Whether al-Qaeda's newer leaders and rank-and-file eventually master the technical intricacies of making and dispersing chemical warfare agents remains an open question.

The state sponsor nexus

One way that terrorists could jump-start their chemical weapons aspirations would be to obtain weapons, materials or technical assistance from states suspected of still harboring chemical weapons programs. A lengthy list of countries was at one time or another in the chemical weapons business, headlined by Russia and the United States, the two largest chemical weapons possessors. Into the early 1990s the US government estimated that over 25 countries were developing and in some cases producing chemical arms.[53] The 1997 activation of the Chemical Weapons Convention (CWC), which bans such activities, greatly reduced the number of states with chemical weapons programs.

Countries that appear on the US government's list of state sponsors of terrorism have repeatedly aided and abetted terrorist activities by funneling money to groups or by assisting or condoning the presence of terrorist training camps on their territory. Table 4.1 crosses the US State Department's 2006 list of state sponsors of terrorism with countries that the US government has named as remaining involved in offensive chemical weapons activities. Other states that have appeared on past lists of suspected chemical weapons proliferators include Egypt, China, India, Indonesia, Iraq and Pakistan.[54] Although none of these states sits on the current US list of terrorist sponsor states, contemporary news accounts reveal that al-Qaeda and affiliated groups and other fundamentalist and separatist groups are active in several of these nations, as well as in Russia. As discussed above, Aum Shinrikyo approached facilities and individuals that the cult believed were affiliated with unconventional weapons programs in Russia and the United States but did not have much, if any, success in soliciting their cooperation.

Table 4.1 State sponsors of terrorism and their chemical weapons activities

US list of terrorist sponsor states (initial listing date)	*Characterization of suspected chemical weapons program*	*Status related to the Chemical Weapons Convention (CWC)*
Iran (1984)	Production of blood, choking agents, mustard gas; R&D on nerve agents	CWC member
North Korea (1988)	Production of blood, choking, nerve agents, mustard gas	Non-signatory
Sudan (1993)	Acquisition of mustard gas from Iraq	CWC member
Syria (1979)	Production of sarin, VX, mustard gas	Non-signatory

Sources: Department of State, Office of the Coordinator for Counterterrorism, *Country Reports on Terrorism 2006* (Washington, DC), April 30, 2007, Chapter 3. Available at: www.state.gov/s/ct/rls/crt/2006/82736.htm. Data on chemical weapons programs of various states can be found at: www.GlobalSecurity.org and cns.miis.edu/research/cbw/possess.htm.

One school of thought holds that states would not assist terrorist groups in acquiring an unconventional weapons capability in part because no state could guarantee that a terrorist group would not turn such weapons back on the state that provided them. Beginning in 1998, however, the US government began voicing concerns about the possible transfer of dangerous technologies, specifically of weapons of mass destruction and their delivery systems, to non-state actors.[55] The fall of the Soviet Union generated worries that unconventional weapons, pertinent materials or technical knowledge might leak from facilities where the scientists and guards were barely being paid. Through the Cooperative Threat Reduction Program and other related programs, the United States led Western nations in efforts to secure and dismantle weapons of mass destruction capabilities in the former Soviet states. These programs to prevent weapons, materials or know-how from falling into the hands of nations or terrorists also included so-called brain drain prevention grants, which employed former Soviet weapons scientists to discourage their possible collaboration with proliferators.[56] While considerable progress has been made in securing and eliminating the former Soviet unconventional arsenal, leakage could still occur if scientists or guards caved to lucrative financial offers or sympathized ideologically with terrorists.

Nor can the possibility that states might give terrorists a helping hand directly be entirely ignored. One Chinese security analyst argued that states that feel threatened by another significantly stronger nation could share unconventional weapons capabilities with terrorists for the purpose of diverting their opponent's attention and resources to the terrorist problem.[57] Given how much state-sponsored assistance or the cooperation of employees at state-run chemical weapons facilities could accelerate terrorist efforts to obtain poison gas, intelligence officials would be well advised to watch vigilantly for any contact between employees of known weapons facilities and terrorists.

Technical advances that could aid terrorist proliferation of chemical weapons

In the past couple of decades, such dramatic and rapid advances have taken place in the revolution of life sciences (e.g. mapping of the human genome, the advent of synthetic biology) that by comparison the field of chemistry appears static. Those who do not look closer may accept the general impression that not much is happening technically in chemistry that would be of interest to aspiring proliferators or concern to agencies of government trying to avert terrorist attacks. While clearly not as dynamic as the life sciences, chemistry is not at a standstill, particularly when it comes to developments of proliferation concern.

This segment of the chapter broaches two such developments, namely the invention of a new generation of more deadly nerve agents and the coming-of-age of the microreactor. The former technical advance is likely to appeal to individuals that act alone rather than to terrorist groups. The latter could be misused by anyone with malicious intent related to chemicals.

Novel warfare agents

Much of the concern about terrorist acquisition and use of unconventional weapons has focused on the intentions of al-Qaeda and other jihadi groups. Terrorist groups, however, are not the only potential malevolent actors. "Lone wolf" actors, individuals with mental problems or perhaps with grievances to settle, have been the source of considerable harm and fear. Perhaps the best-known solo terrorist of modern times is Ted Kaczinski, the "unabomber," who eluded the Federal Bureau of Investigation while sending 23 bombs to targets from 1978 until his capture in early April 1996.[58] A lone actor with sufficient technical skills and a desire to make some sort of out-of-the-ordinary statement via the type of weapon used for attacks might consider poison gas to be an enviable weapon. Ted Kubergovic, tagged as the Alphabet Bomber, was such an individual. Among the materials that police found in his home in August 1974 was 20 lbs of sodium cyanide, which can be used to release hydrogen cyanide gas or to make the nerve agent tabun.[59]

Some lone actors might look beyond the ordinary list of chemical warfare nasties (e.g. mustard gas, VX) for next-generation warfare agents in order to demonstrate above-average technical prowess and to create extra alarm and harm. Such individuals are likely to zero-in quickly on Dr. Vil Mirzayanov, the 26-year veteran of the Soviet chemical weapons complex who blew the whistle on the *novichok* program in 1991.[60] *Novichok*, Russian for "newcomer," refers to a group of binary warfare agents wherein for safety reasons the final chemical components are stored separately and combined just prior to use. Mirzayanov stated that Soviet test results proved the *novichok* agents to be far more lethal than the nerve agents VX and soman. The concept behind the *novichok* program was to bury a latent chemical weapons production capability within the agro-chemical industry. Though Mirzayanov himself did not reveal the formulas for the *novichok* agents, sufficient relevant information has appeared in the public domain to embolden proliferators to replicate these agents. Moreover, Czech scientists have described a second class of next-generation nerve agents in a few scientific articles. One of the agents in this new family of poison gas combines the characteristics of VX and sarin to enhance lethality via both the skin and inhalational exposure routes.[61]

Whereas terrorist groups tend to be expedient and go for proven, readily acquired weapons, lone actors may relish the technical challenge of making a novel nerve agent. Deranged individuals may also like the idea that their attack(s) would have the hallmark of a previously unused poison gas. Once such individuals have dissected the literature for the chemical ingredients required for these next-generation warfare agents, the chemicals may be available without the controls that normally apply to the precursors for the classic warfare agents because the CWC does not specifically ban the *novichok* and Czech agents.[62] The indicators that a lone terrorist is dabbling in novel chemical warfare agents are likely to be very discrete because their production capacity is also likely to be small. Should lone actors gravitate toward novel chemical warfare agents, even small-scale attacks with these agents could bring about considerable harm and disruption.

Microreactors

Traditionally, chemical weapons programs have necessitated a fairly large infrastructure to manufacture large amounts of precursor chemicals and then the warfare agents themselves, not to mention to fill and store militarily significant quantities of munitions and bulk agent. One of the major tip-offs to the existence of an illicit chemical weapons program has been large-scale chemical facilities, whether in dedicated military production sites or in commercial plants masquerading as manufacturers of fertilizers and pesticides. Recent advances in the miniaturization of chemical production technologies could, however, vastly reduce the visible footprint of chemical weapons activities.[63]

Microtechnology was first applied to chemical reactors in the 1970s, but it was not until the late 1990s that the development and use of microreactors began to flourish. The most striking feature of a microreactor is its compactness: the dimensions of a microreactor can range from the size of a coin or a credit card to the size of a notebook.[64] The inner channels of microreactors are usually under 1 mm in diameter and often etched in glass.[65] These inner channels, of which there can be hundreds in a single microreactor, create a high surface-area-to-reactant ratio that solves one of the problems that has plagued conventional chemical reactors. The much greater internal surface volume of a microreactor allows the liquid to be distributed throughout these channels, and the closer proximity of the heat exchangers to these channels enables more precise regulation of the heat that chemical reactions generate. Since microreactors dissipate heat more quickly than their ordinary counterparts, they are better suited for reactions that are highly exothermic and potentially explosive. In addition, the small size of microreactors permits much tighter control over agitation, pressure, flow, and other reaction conditions than is possible with standard industrial reactors. Reactions are also much faster because one-molecule-at-a-time simple diffusion replaces the relative inefficiency of mixing reactants using big propellers inside of a large tank. This instantaneous mixing speeds up production, which in turn reduces production costs.[66]

The features of microreactors mean that new chemical processes that would previously have been difficult or even impossible are now technically feasible. Accordingly, microreactors are being exploited in the civilian sector with growing depth and scope of purpose, including for the manufacture of drugs for clinical trials.[67] To increase yields to industrial scale, microreactors can be connected in parallel arrays to process either identical or different reactions, producing the same output as an ordinary chemical plant in less time, more safely and with a higher quality product.[68] Industrial scale quantities of product – tons as opposed to gallons – can be generated by increasing the number of microreactor arrays involved.

Microreactors have already been used to synthesize numerous extremely hazardous chemicals, including hydrogen cyanide and methyl isocyanate.[69] The latter, an unstable and highly toxic substance, is infamous for causing an estimated 3,800 initial and at least 15,000 subsequent deaths when it was released in 1984 from a Union Carbide plant in Bhopal, India.[70] Furthermore, Hastelloy microreactors are under development and their introduction will facilitate the manufacture of extremely corrosive chemicals – including organophosphate chemicals used to make pesticides, fertilizers and nerve agents – in microreactors.

For terrorists seeking an in-house chemical weapons production capability, the implications of microreactor technology are quite startling. Aum Shinrikyo encountered technical problems with the scale-up of its chemical weapons manufacturing process from beakers to reactors. Had Aum Shinrikyo's scientists turned to microreactors, they might have averted several technical obstacles and reached their goal of producing 70 tons of sarin in 40 days.

Terrorists could employ microreactors to skirt export controls and make their own key chemical weapons precursors, such as thionyl chloride or methylphosphonyl difluoride. Though working from slightly different control lists, the export control cooperative called the Australia Group and the CWC restrict trade in these and other high proliferation-risk chemicals, hindering the ability of proliferators to obtain them readily.[71] Proliferators must instead resort to the procurement of more basic chemicals and engage in some reasonably complicated domestic production of key precursors to be able to make the classic chemical warfare agents.[72] While the production of chemical agents has a reputation as being "easy," it would be more accurate to say that chemical weapons proliferation is easier than the proliferation of biological or nuclear weapons, not simply easy in and of itself. To illustrate, Libya's chemical weapons program floundered largely because the Libyans were unable to scale-up the manufacture of methylphosphonyl difluoride. Thus, tapping microreactor technology to make key chemical precursors, not to mention the warfare agents themselves, could give proliferators a considerable advantage and has the added attraction of stripping those attempting to monitor proliferation of yet another tip-off to their illicit activity.

Terrorists in search of knowledge about microreactors could consult the more than 1,000 recently published studies about possible uses of the technology.[73] Or, they could attend academic and industrial meetings where microreactors are routinely discussed.[74] The cost of the most-purchased microreactors range between $70,000 and $200,000. Over a dozen companies are developing and producing microreactors.[75] Microreactors are not currently on the Australia Group's list of export control items, leaving the marketplace quite accessible to terrorists.

In sum, these small devices require a fraction of the space of an ordinary chemical plant and yet can churn out large volumes of ultra-toxic chemicals faster, more safely, on demand and more economically. This technology could obviate some of the major customary signatures of a chemical weapons program, rendering both state- and terrorist-run chemical weapons programs virtually undetectable. Microreactors could therefore beckon terrorists who seek to expand their arsenal in this direction.

Short cuts to chemical disaster

For terrorists aiming to cause a chemical disaster, arranging to get chemical weapons or materials from known possessors or on the black market or assembling from scratch a chemical weapons production capability are relatively complicated endeavors that carry a certain amount of risk of detection and capture. A less risky route to the same outcome involves foul play with industrial chemical facilities and chemicals in transit. Industrial chemicals with mass-casualty potential are common in developed nations. To illustrate, in 1999 the Environmental Protection Agency put the number of US facilities working with hazardous or extremely hazardous chemicals at approximately 850,000.[76] Of note, one trademark of al-Qaeda has been to employ the infrastructure of modern society in its attacks.

The chemical industry has traditionally concentrated in the northern hemisphere, but in response to the rising price of feedstock chemicals, labor and energy costs and tougher environmental regulations in Western countries, the industry has shown marked growth in China, India, South Korea, Nigeria, Trinidad, Thailand, Brazil, Venezuela, Indonesia, Southeast Asia and the Middle East. The global spread of the chemical industry has several implications for terrorists. For starters, terrorists have increasing opportunities to get hired as menial or technical workers as more chemical plants crop up in far-flung countries. Once employed in these plants, such individuals can get hands-on operational training, including in the handling and manufacture of corrosive, toxic chemicals and in the other industry operational practices. The former type of knowledge would help terrorists overcome the technical hurdles involved in scaling-up production of hazardous chemicals to significant quantities; the latter in exploiting the security shortcomings of industrial chemical sites.

The security weaknesses of chemical industry facilities are not difficult to identify. A chemical industry expert notes that site security is the responsibility of facility operators who are first and foremost concerned with meeting health, safety and environmental regulations. While chemical plants probably have fences and gates, the primary concern of any guard(s) that may be on-site is to check the identity of visitors and see that they get the requisite site safety briefing. For safety reasons, plant guards are not armed. Storage tanks will have safety markings (e.g. degree of flammability of contents) but some multinational chemical companies have policies that otherwise instruct operators to label tanks falsely to prevent competitors from learning their reactants. In some countries, plant operators do not bother to mark tanks at all. As for internal security when hiring, chemical companies check the technical credentials of applicants and may call references, but they do not investigate the background of incoming employees from a security perspective.[77] These circumstances render chemical plants relatively easy to infiltrate from within or to attack from outside.

In transit, the security for chemicals is just as lacking. Shippers provide notice to port authorities so that they know the type and quantity of inbound cargo that has to be unloaded. No special security precautions are taken for chemical shipments, leaving them easy to scout for targeting. One-ton chemical containers are shipped by rail, and in many countries these containers are required for safety reasons to have basic markings indicating the flammability and stability of the container's contents. In rail yards, these labels can be read to provide terrorists with a first-order categorization of containers to target. From a minor initial detonation or rupturing of rail or sea cargo shipment containers, the catalytic properties of chemicals located in close proximity to each other could create a toxic soup.[78]

A great many of the chemical industry's facilities are located in heavily populated areas, and security analysts have argued that such sites therefore make tempting targets for terrorists.[79] The scale of possible harm from sabotage of chemical facilities or from shipments in ports or rail yards located in metropolitan areas is noteworthy. According to data from 2000, a worst-case scenario event at 123 facilities in the United States would leave over one million people at risk of death or injury. At another 586 US chemical plants, a worst-case event would place between 100,000 and one million people at similar risk.[80] Another characterization of the scope of the problem of possible sabotage of chemical industry facilities is that according to one industry veteran, roughly 550–600 industry facilities globally are engaged in the production, processing and consumption of the approximately 50 or so industrial chemicals that could seriously jeopardize human life (e.g. methyl isocyanate, phosgene).[81]

Given the minimal security for hazardous commercial chemicals, both *in situ* and in transit, these facilities and shipments could summon the attention of terrorist groups or individuals with chemical mayhem in mind. Modest operational planning is needed, and the only visible tip-off to forthcoming attacks may be the perpetrators' reconnaissance of targeted chemical plants or transportation nodes. For al-Qaeda and affiliated groups, part of the appeal of sabotaging chemical plants or transportation centers is the indiscriminate harm that will result. The resulting toxic cloud will blow with the wind, leaving Allah to decide who lives, is injured or dies.

Concluding observations

Virtually every activity carries with it indicators that would allow outsiders to see that something is underway. Even for clandestine activities, front companies will materialize, meetings will be held, materials will be procured, buildings will be bought or rented to manufacture and store items. Once a terrorist group forms, these types of activities can be crossed with such factors as the group's ideology and tactics to assist in a determination of whether the group is after unconventional weapons.

As law enforcement and intelligence officials struggle to distinguish individual activities and patterns significant to the pursuit of chemical weapons from the vast array of available and collected data, they need to take note of anomalies. A news or raw intelligence report of a strange activity that does not seem to fit any context or recognized pattern should command some attention and reflection. Terrorists, like all other humans, make mistakes; anomalies can be the visible tip of such mistakes, the key to understanding the behavior of a terrorist group. To wit, why did moon-suited Aum Shinrikyo members twice try to blow foul-smelling material off the roof of the cult's headquarters building in the summer of 1994? For that matter, what was causing the awful smell that neighbors reported coming from the cult's Kamikuishiki compound in the fall of 1994? The answers to these questions contained information important to unraveling Aum Shinrikyo's unconventional weapons activities and intentions. Otherwise, these activities were just oddities. Along with several other categories of identifiers, Table 4.2 lists therefore lists anomalous activity as something that might help intelligence and law enforcement authorities penetrate the intentions and activities of terrorist groups.

Table 4.2 Indicators of possible terrorist pursuit of chemical weapons

Category	*Detail(s)*
Ideology	• Chemical weapons stipulated in or consistent with ideology • Weapons of mass destruction stipulated in or consistent with ideology
Tactics	• Attacks intended to cause mass casualties • Attacks that employ unconventional or novel weaponry
Propaganda	• Specific mention of chemical weapons in statements, speeches, videos, pamphlets, and in other written and visual material • Reference(s) to other unconventional or unusual weapons
Recruitment practices	• Emphasis on recruiting technically skilled individuals (e.g. chemistry, engineering) • Recruitment activities (e.g. meetings, fliers, other propaganda)
Training materials and activities	• Dedicated instruction in training manuals, videos and/or in hands-on training activities related to manufacture of, use of and/or protection against chemical weapons • Similar dedicated instruction pertaining to other unconventional or unusual weapons
Interaction(s) with state-run chemical weapons facilities	• Attempts to contact government facilities engaged in offensive or defensive chemical weapons activities, particularly in states known for facilitating other types of terrorist activity
Efforts to acquire other unconventional weapons	• Efforts to obtain biological, radiological or nuclear weaponry on the black market or through procurement of needed equipment, materials and know-how
Unusual features at facility	• Above-average physical security (e.g. barriers, guards) possibly denotes high-value personnel and/or activity • Pollution abatement equipment, bermed building(s) and/or very remote location possibly denotes research, testing, processing or storage of hazardous materials

Purchase of basic or direct precursor chemicals	• Purchase of large quantities of building-block chemicals or smaller amounts of direct precursors, particularly of multiple purchases from multiple supply companies • Use of front companies for purchases

Purchase of specialized equipment or materials	• Acquisition of a large quantity of a specific item or the collective purchase of multiple items:
	  • Hastelloy or glass-lined reactors and piping
	  • Analytical equipment (e.g. gas chromatograph/mass spectrometer)
	  • Pollution abatement equipment
	  • Safety equipment:
	    • Personal protective equipment
	    • Decontamination equipment
	    • Antidotes
	• Use of front-companies for purchases

Purchase pattern(s) inconsistent with output	• Absence of commercial product(s) made from purchased equipment and chemicals, including the lack of advertisement, production and distribution • Implausible cover stories for purported commercial sites with little or no output
Dress rehearsals	• Weapons field tests • Assassination attempts using a particular weapon or tactic • Small-scale attacks to test a weapon or tactic
Reconnaissance	• Scoping a building or location to determine the security (e.g. fences, guards) or physical features (e.g. ventilation system, traffic choke points) that need to be taken into account to execute an attack
Anomalies	• Unusual, perplexing activity involving suspected terrorists or terrorist facilities

Regrettably, for every indicator of a possible covert chemical weapons activity, a wily and determined terrorist group or lone actor can probably devise an alternative way to accomplish the activity or to disguise it to hinder discovery. As noted above, other factors could make detection more difficult, including the possible terrorist use of microreactors to produce warfare agents, precursors or other hazardous chemicals. Also, unlike governments engaged in proliferation, terrorists do not require "militarily significant" quantities of chemical agent to saturate a battlefield, especially if their goal is to petrify people, not kill massive crowds. Moreover, terrorists may not be preoccupied with Western safety standards; groups that place little emphasis on the survival of the foot soldiers who carry out attacks may not bother to buy and install equipment that protects workers or the public nearby from accidental exposure to harmful substances. The likely absence of some of the telltale characteristics of a state-sponsored chemical weapons program (e.g. exhaust scrubbers) in terrorist proliferation activities will only add to the complexity of the task facing intelligence and law enforcement agencies.

A final indicator that could influence the behavior of terrorists is what might be called the harbinger event – the attack so successful, so simple yet insidious, so accessible, or so spectacular that it inspires countless others to imitate it. Many security analysts assessed Aum Shinrikyo's 1995 subway sarin attack to be such a watershed event for unconventional terrorism, but over a decade later a horde of chemical copycats has yet to appear. Though headline grabbing, perhaps other terrorists viewed Aum's quest for and use of sarin as more trouble than it was worth, especially since conventional bombs are far easier to acquire and could produce death and injury on a similar scale. No one can be sure why Aum's sarin attack did not spark copycats, yet suicide bombing caught on with the Tamil Tigers; Hezbollah in Lebanon; Sikh and Kurdish nationalists in India and Turkey, respectively; Chechen separatists in Russia; and Hamas, Islamic Jihad and al-Qaeda, the latter despite the Islamic prohibition against suicide.

In 2007, a series of over 15 attacks in Iraq combined explosives with the commercial chemical chlorine, and again raised the prospect that terrorists might more routinely use chemicals as an attack tool. Intense exposure to chlorine, used frequently on World War I battlefields, can be lethal, but chlorine is not well-suited to cause extensive harm when married with explosives. In the first attack on January 28, 2007 the insurgents may have mistaken a tank of chlorine for a tank containing an accelerant (e.g. propane) that would have boosted the conventional explosion, or they may have purposefully sought to exacerbate injuries and fear by dispersing chlorine, a yellow gas. Regardless, the attack killed 16, reportedly from blast effects. The ready availability of chlorine (e.g. interception of tanker convoys, theft from water treatment plants) may have contributed to the insurgents' decision to continue its use. The insurgents' proficiency in dispersing the chlorine gas increased as they altered the explosive composition of their bombs. In one of three separate chlorine attacks on March 16, 2007 some 250 suffered the effects of chlorine exposure (e.g. burns, respiratory difficulty, nausea).[82]

Since most of the injuries from chlorine in the 2007 attacks were not life threatening, the primary impact of these attacks, supposedly the work of the "takfiri" or al-Qaeda in Iraq, was to exacerbate the panic among Iraqis about the threat of poisoning. Though the use of chlorine dissipated in mid-2007, this series of attacks established proof of principle that commercial chemicals can be effectively integrated into the terrorist toolkit. Iraq has become something of a training ground for al-Qaeda and its affiliated organizations, so the chlorine attacks of 2007 raised again the question of whether these chemical attacks would blossom into more widespread use. Terrorists could take any number of lessons from this chlorine attack campaign, including the thought of incorporating different or more hazardous commercial chemicals into attacks in Iraq or of taking commercial chemical attacks into other nations.

A transition to more frequent terrorist use of chemicals is something that the passage of time will reveal. For terrorists bent on crossing the threshold to the use of unconventional weapons, the barriers related to technical knowledge, materials and equipment are arguably lower for chemical arms than for nuclear and even biological weaponry. In the scheme of chemical terrorism possibilities, small-scale chemical attacks or larger toxic disasters engineered through sabotage of industrial plants or of hazardous chemicals in transit are the most readily achievable attack options and therefore the likely indicators that such a transition may be in the offing.

Notes

1 From 1997–2006, 5,798 armed attacks took place. RAND/MIPT Terrorism Knowledge Base, available at: www.mipt.org. This database is one of the most comprehensive sources of information available on domestic and international terrorist activities over the past 35 years.

2 See for example, Richard A. Falkenrath, Robert D. Newman and Bradley A. Thayer, *America's Achilles' Heel: Nuclear Biological, and Chemical Terrorism and Covert Attack* (Cambridge, MA: MIT Press, 1998); Walter Laquer, "Postmodern Terrorism," *Foreign Affairs* 75:5 (1996): 23–36.

3 For an initial report of the incident see, Stephanie Storm, "Possible Cyanide Poisoning at Festival Startles Japan," *New York Times*, July 29, 1998. Available at: www.newyorktimes.com. In addition to the RAND/MIPT Terrorism Knowledge Base, information about unconventional terrorist attacks can also be found in the database of the James Martin Center for Nonproliferation Studies, available at: www.nti.org/db/cbw/index.htm.

4 For the medical effects of chemical agents see, Frederick R. Sidell, Ernest T. Takafuji and David R. Franz (eds.) *Textbook of Military Medicine: Medical Aspects of Chemical and Biological Warfare, Part I: Warfare, Weaponry, and the Casualty* (Washington, DC: Office of the Surgeon General, US Department of the Army, 1977). More briefly, Adrian Dwyer, John Eldridge and Mick Kernan, *Jane's Chem-Bio Handbook*, 2nd edn (Surrey, UK: Jane's Information Group, 2002).

5 Stephen Preisler, *Silent Death*, 2nd edn (Green Bay, WI: Festering Publications, 1997). For another example of this type of cookbook, Maxwell Hutchkinson, *The Poisoner's Handbook* (Port Townsend, WA: Loompanics Unlimited, 1988).

6 For estimates of the skill level required see, Ron Purver, *Chemical and Biological Terrorism: The Threat According to the Open Literature* (Ottawa: Canadian Security Intelligence Service, 1994), p. 69; Falkenrath *et al.*, *America's Achilles Heel*, pp. 102, 106.

7 A mustard production plant could cost $5 million to $10 million, a nerve agent production facility about $20 million. Office of Technology Assessment, *Technologies Underlying Weapons of Mass Destruction*, OTA-BP-ISC-115 (Washington, DC: US Government Printing Office, 1993), p. 27.

8 After the Matsumoto sarin attack, which killed seven citizens, Japanese police strongly suspected the cult was making sarin inside its Kamikuishiki compound. The insider tip came from two cult members who were in Japan's Self Defense Force, where police sent 500 officers for last-minute chemical defense training prior to the raid. D.W. Brackett, *Holy Terror: Armageddon in Tokyo* (New York: Weatherhill, 1996), pp. 121–4, 126–7; David E. Kaplan and Andrew Marshall, *The Cult at the End of the World* (New York: Crown Publishers, Inc., 1996), pp. 237–8; Anthony T. Tu, "Aum Shinrikyo's Chemical and Biological Weapons," *Archives of Toxicology, Kinetics and Xenobiotic Metabolism* 7:3 (1999): 55, 57; Kyle B. Olson, "Aum Shinrikyo: Once and Future Threat?" *Emerging Infectious Diseases* 5:4 (1999): 42–7; Kyle B. Olson, "Overview: Recent Incident and Responder Implications," in Office of Emergency Preparedness, *Proceedings of the Seminar on Responding to the Consequences of Chemical and Biological Terrorism* (Washington, DC: US Public Health Service, Department of Health and Human Services, July 4, 1995), pp. 2–38.

9 On the casualty statistics from the attack see, Dr. Fred Sidell, "U.S. Medical Team Briefing," in *Proceedings of the Seminar on Responding to the Consequences of Chemical and Biological Terrorism*, pp. 2–33.

10 The cult's followers were told that after victory in the cataclysmic battle, they would remain as the super race. Kaplan and Marshall, *The Cult at the End of the World*, pp. 15–17, 46–9, 84–5, 222–3, 270; Brackett, *Holy Terror*, 81–4, 95–6, 105, 121–4; Murray Sayle, "Nerve Gas and the Four Noble Truths," *New Yorker* (April 1, 1996): 62–3; Tu, "Aum Shinrikyo's Chemical and Biological Weapons," 46, 49.

11 Sayle, "Nerve Gas and the Four Noble Truths," 68. See also, Brackett, *Holy Terror*, pp. 102, 105–6.

12 Kaplan and Marshall, *The Cult at the End of the World*, pp. 87–8, 102, 213–14, 216.

13 Aum has been erroneously credited with dispersing anthrax and botulinum toxin on several occasions, but the cult's strains of these agents were harmless and its dispersal systems totally ineffective. For more on the cult's biological weapons activities see, Amy E. Smithson and Leslie Anne Levy, *Ataxia: The Chemical and Biological Terrorism Threat and the US Response* (Washington, DC: Henry L. Stimson Center, 2000), pp. 75–80.

14 Theoretically, 250 tons of sarin would be over 1.25 billion lethal doses, but that death toll would result only if the dispersal of the agent managed to deliver the requisite amount to all 1.2 billion people. Prepared Statement of Connie J. Fenchel, US Congress, Senate Committee on Governmental Affairs, Permanent Subcommittee on Investigations, *Global Proliferation of Weapons of Mass Destruction*, 104th Cong., 1st sess., October 31, 1995 and November 1, 1995 (Washington, DC: US Government Printing Office, 1996), p. 229.

15 Author's interview with an individual who worked at Tooele Depot at the time this incident occurred, July 17, 2007, Washington, DC.

16 Staff Statement [and testimony of US intelligence officials], *Global Proliferation of Weapons of Mass Destruction*.

17 Kaplan and Marshall, *The Cult at the End of the World*, pp. 107–8, 119–20, 216, 259; David E. Kaplan, "Aum Shinrikyo (1995)," in Jonathan B. Tucker (ed.) *Toxic Terror: Assessing the Terrorist Use of Chemical and Biological Weapons* (Boston, Mass: MIT Press, 2000), pp. 293–5.

18 Japanese police found the GSP-11 gas detector at Kamikuishiki. Prepared Statement of Gordon C. Oehler, *Global Proliferation of Weapons of Mass Destruction*, p. 216.

19 Kaplan and Marshall, *The Cult at the End of the World*, pp. 87–8, 102, 213–14, 216.

20 Each of the laboratories contained sophisticated equipment, such as $10,000 infrared spectrometers to separate and analyze chemical compounds. The facility was so advanced that it had a fluorine treatment capacity, a feature that numerous Western laboratories bypass because of its expense. The individual who reportedly sold Aum the blueprints for the plant was the Secretary of the Russian Security Council, Oleg Lobov. Kaplan and Marshall, *The Cult at the End of the World*, pp. 119–20, 216, 259; Kaplan, "Aum Shinrikyo (1995)," pp. 293–5. See also, Olson, "Overview," pp. 2–42; Tu, "Aum Shinrikyo's Chemical and Biological Weapons," 48, 79.

21 Testimony of John F. Sopko, *Global Proliferation of Weapons of Mass Destruction*, p. 21; Kaplan and Marshall, *The Cult at the End of the World*, pp. 120–1, 149. See also, Anthony T. Tu, *Chemical Terrorism: Horrors in Tokyo Subway and Matsumoto City* (Fort Collins, CO: Alaken, Inc., 2002).

22 Tsuchiya stated that one of the reasons he signed on with the cult was that Aum's laboratories were better than those at Tsukuba University. Sayle, "Nerve Gas and the Four Noble Truths," 70. Also, Staff Statement, *Global Proliferation of Weapons of Mass Destruction*, p. 61; Brackett, *Holy Terror*, pp. 114–15. Tsuchiya has been described as a "brilliant" chemist. Tu, "Aum Shinrikyo's Chemical and Biological Weapons," 48, 50.

23 Anthony T. Tu, "Basic Information on Nerve Gas and the Use of Sarin by Aum Shinrikyo," *Journal of the Mass Spectrometry Society of Japan* 44:3 (1996): 311.

Public sources differ on the amount of chemicals the cult had. See also, Kaplan and Marshall, *The Cult at the End of the World*, pp. 98, 257.

24 Nerve antidote was administered to Aum's chief of security, Tomomitsu Miimi, when an attempt to attack a rival cult with sarin backfired. The subway attackers on March 20, 1995 also took the antidote before the mission and carried additional antidote with them. When Satyan 7 was in full-swing operation, workers wore gas masks and full-body suits for such activities as sampling. Anthony T. Tu, "Chemistry and Toxicology of Nerve Gas Incidents in Japan in 1994 and 1995," paper presented at the Third Congress of Toxicology in Developing Countries, September 1, 1996, Cairo; Kaplan and Marshall, *The Cult at the End of the World*, pp. 120, 131–3, 150, 259.

25 Unlike their chemical warfare agents, Aum's biological substances were harmless since no laboratory rats died when exposed to their *C. botulinum*. Testimony of Kyle B. Olson, *Global Proliferation of Weapons of Mass Destruction*, p. 111; Kaplan and Marshall, *The Cult at the End of the World*, p. 97.

26 Kaplan and Marshall, *The Cult at the End of the World*, pp. 127–9, 133; The Australian Investigation of the Aum Shinrikyo Sect, Staff Statement and Testimony of Alan Edelman, in *Global Proliferation of Weapons of Mass Destruction*, pp. 32, 75, 613. The cult apparently intended to mine uranium ore, a base material for nuclear weapons, from this site. Brackett, *Holy Terror*, pp. 93–4, 114.

27 Tu, "Chemistry and Toxicology of Nerve Gas Incidents in Japan in 1994 and 1995"; Kaplan and Marshall, *The Cult at the End of the World*, pp. 131–3. Another version of this attack describes only one attempt, during which the dispersal system caught fire. Brackett, *Holy Terror*, p. 29. On the backwards spraying in the first attempt, Dr. Anthony Tu, telephone interview with author, October 6, 2000.

28 Pamela Zurer, "Japanese Cult Used VX to Slay Member," *Chemical & Engineering News* (August 31, 1998): 7; Tu, "Aum Shinrikyo's Chemical and Biological Weapons," 75, 79.

29 Kaplan and Marshall, *The Cult at the End of the World*, pp. 138–9; John F. Sopko, "The Changing Proliferation Threat," *Foreign Policy* 105 (1996): 13. The original attack plan called for the cult to spray sarin into the courthouse where the judges worked in broad daylight. Brackett, *Holy Terror*, pp. 27–8.

30 Japanese prosecutors estimated the amount of sarin released. Testimony of John F. Sopko, *Global Proliferation of Weapons of Mass Destruction*, p. 21. Kaplan and Marshall, *The Cult at the End of the World*, p. 144.

31 Authorities positively identified sarin using gas chromatography, mass spectroscopy, chemical ionization and other methods. Traces of sarin, phosphoric acid and di-isopropyl were also found in the dead victims. For more on the patients' symptoms and treatment, see Dr. Nobu Yanagisawa, "Matsumoto, Japan (June 1994)," in *Proceedings of the Seminar on Responding to the Consequences of Chemical and Biological Terrorism*, pp. 2–13–18, 2–20; Hiroshi Okudera, H. Morita, T. Iwashita, T. Shibata, T. Otagiri, S. Kobayashi, N. Yanagisawa, "Unexpected Nerve Gas Exposure in the City of Matsumoto: Report of Rescue Activity in the First Sarin Gas Terrorism," *American Journal of Emergency Medicine* 15:5 (1997): 528. Note that the casualty figures in these accounts differ slightly, and a survey later revealed that another 277 people experienced symptoms of sarin exposure but did not seek medical assistance. The seven who died were located in three buildings in close proximity to the parking lot where the sarin was released. Brackett, *Holy Terror*, p. 37.

32 Kaplan and Marshall, *The Cult at the End of the World*, pp. 144–6.

33 Tu, telephone interview with author, October 6, 2000. The cult's scientific staff may have been book-smart, but they lacked operational experience, as shown by the leakage of toxic chemicals outside of the Kamikuishiki fence line and worker injuries. Satyan 7 was equipped with sealed rooms, ventilation and a decontamination chamber. Workers were required to wear gas masks and full body suits during high-risk operations but on numerous occasions Satyan 7 personnel were still exposed to

toxic chemicals, experiencing symptoms ranging from nosebleeds to convulsions. Brackett, *Holy Terror*, p. 116.

34 Milton Leitenberg, "Aum Shinrikyo's Efforts to Produce Biological Weapons: A Case Study in the Serial Propagation of Misinformation," *Terrorism and Political Violence* 11:4 (1999): 152, fn. 6. Also on this attempted attack, Sheryl WuDunn, Judith Miller and William J. Broad, "Sowing Death: How Japan Germ Terror Alerted the World," *New York Times* (May 26, 1998); Staff Statement, *Global Proliferation of Weapons of Mass Destruction*, p. 63; Kaplan and Marshall, *The Cult at the End of the World*, pp. 94–6. Some accounts said that birds were killed and pets were sickened, but while some people lost their appetites, no one died.

35 Trial testimony of cult member Shigeo Sugimoto as reported in March 19, 1997 edition of *Asahi Shinbun*. Excerpt provided by Mr. Masaaki Sugishima, September 1, 2000; WuDunn *et al.*, "Sowing Death;" Kaplan and Marshall, *The Cult at the End of the World*, p. 57.

36 WuDunn *et al.*, "Sowing Death"; Kaplan and Marshall, *The Cult at the End of the World*, pp. 235–6; Staff Statement, *Global Proliferation of Weapons of Mass Destruction*, p. 63; Olson, "Overview," pp. 2–37, 2–59.

37 Kaplan and Marshall, *The Cult at the End of the World*, pp. 241–2; Tu, "Aum Shinrikyo's Chemical and Biological Weapons," 55.

38 Through its fiery rhetoric and actions, al-Qaeda has perhaps reshaped the Qur'an's jihad concept about the removal of persecution and the establishment of the supremacy of Islam. The term jihad, as used in passages of the Qur'an, has various meanings, ranging from a non-violent, peaceful struggle to the conquest of one's own evil desires to an armed conflict. The meaning of jihad is widely discussed in the literature on Islam and terrorism. Briefly, Douglas E. Streusand, "What Does Jihad Mean?" *Middle East Studies Quarterly* IV:3 (1997). Available at: www.meforum.org/article/357. More at length, Reuven Firestone, *Jihad: The Origins of Holy War in Islam* (Oxford: Oxford University Press, 1997).

39 The English translation of Bin Laden's 1996 fatwa can be read at: www.pbs.org/newshour/terrorism/international/fatwa_1996.html.

40 In the same interview with *Time* magazine, bin Laden also noted that "hostility against America" was a "religious duty." In response to a question asserting that al-Qaeda was attempting to get chemical and nuclear weapons, bin Laden said it was "a sin" not to try. Rahimullah Yusufzai, "Conversation with Terror," *Time* 153:1 (January 11, 1999). Furthermore, in an interview aired on December 24, 1998, when ABC News correspondent John Miller asked bin Laden whether he was seeking to obtain chemical or nuclear weapons, bin Laden replied, "Acquiring weapons for the defense of Muslims is a religious duty. If I have indeed acquired these weapons, then I thank God for enabling me to do so." The transcript from this interview can be found at: www.ABCNEWS.com.

41 Al-Fahd tallies the number of Muslim faithful purportedly killed over the years as a result of US foreign and defense policies, coming up with a total of ten million, and argues that this justifies the use of unconventional weapons in response. For an insightful analysis of Al-Fahd's fatwa, see Reuven Paz, "YES to WMD: The first Islamist Fatwah on the use of Weapons of Mass Destruction," *PRISM Series of Special Dispatches on Global Jihad* 1. Available at: www.e-prism.org/projectsandproducts.html. Also available at this website is Paz's more recent "Global Jihad and WMD: Between Martyrdom and Mass Destruction," released by the Hudson Institute in September 2005.

42 See interview excerpts from the Pakistani newspaper *Dawn* in Tim Weiner, "Bin Laden Asserts He has Nuclear Arms," *New York Times*, (November 10, 2001).

43 For an analysis of the online material and discussion of unconventional weaponry at websites frequented by jihadists and their sympathizers, see Anne Stenerson's chapter in this volume.

44 *Report to the President of the United States, Commission on the Intelligence Capabilities of the United States Regarding Weapons of Mass Destruction* (Washington, DC: US General Accounting Office, March 31, 2005), pp. 270–1.

45 The computer that contained a record of the Curdled Milk program belonged to top al-Qaeda deputy Ayman al-Zawahiri. A chemical engineer, Abu Khabbab, worked in this weapons program. The computer in question was recovered when a reporter bought it in Kabul. Alan Culluson and Andrew Higgins, "Computer in Kabul Holds Chilling Memos," *Wall Street Journal* (December 31, 2001); "Report: Al Qaeda Computer Had Plans for Bio-Weapons," Reuters, December 21, 2001. Other documents related to the program were recovered with the capture of Khalid Sheik Muhammad. Barton Gellman, "Al-Qaida Near Biological, Chemical Arms Production," *Washington Post* (March 23, 2003).

46 This video, apparently one of 64 tapes, is part of al-Qaeda's video archive of its activities. Nic Robertson, "Tapes Shed New Light on Bin Laden's Network," CNN, August 19, 2002. Available at: http://archives.cnn.com/2002/US/08/18/terror.tape.main. The CNN tape shows the death of three dogs. In one scene, several individuals in Afghani style sandals are told to leave a room containing a dog. John Gilbert, Jonathan Tucker and David Kay attribute the dog's symptoms to those sarin or cyanide. According to expert Dr. Fred Sidell, the dog does not react terribly quickly to the white vapor, as would be the case with nerve agent. Sidell also observes that the dog experiences selective paralysis of hind quarters, which is not consistent with exposure to cyanide or sarin. In testimony, a former al-Qaeda member says the chemical involved was cyanide, which was also used in tests with rabbits. The transcript of this CNN program, "Insight," which aired on August 19, 2002, can be found at: http://edition.cnn.com/TRANSCRIPTS/0208/19/i_ins.01.html.

47 The US government also believed al-Qaeda had a modest biological weapons capability. In Senate testimony, former Central Intelligence Agency Director George Tenet noted Bin Laden's interest in chemical weapons and the group's training for attacks with chemical and biological agents. Pamela Hess, "Al Qaida May Have Chemical Weapons," United Press International, August 19, 2002; Judith Miller, "Lab Suggests Qaeda Planned to Build Arms, Officials Say," *New York Times* (September 14, 2002).

48 The chemical in question is highly lethal but apparently would not have made the most effective dirty bomb. For an analysis of this plot, see: http://cns.miis.edu/pubs/week/040413.htm.

49 Secondary targets for the attack were the Jordanian Prime Minister's office and the US embassy. The plot was reportedly hatched on the order of the top of al-Qaeda in Iraq at the time, Abu Musab al-Zarkawi, who Jordanian officials sentenced to death in absentia and who was later killed in Iraq. Charles Hanley, "WMD Terrorism: Sum of All Fears Doesn't Always Add up," *USA Today* (October 29, 2006). Available at: www.usatoday.com/news/world/2005-10-29-terror-vision_x.htm. See also, "Jordan Sentences Zarkawi, in Absentia, for Chemical Plot," Associated Press, in *New York Times* (February 16, 2006). Available at: www.nytimes.com/2006/02/16/international/middleeast/16amman.html Also, "Jordon Says Major Al Qaeda Plot Disrupted," April 26, 2004. Available at: www.cnn.com/2004/WORLD/meast/04/26/jordan.terror.

50 Al-Masri's death occurred in a mid-January 2006 air strike in Damadola, Pakistan. Habibullah Khan and Brian Ross, "U.S. Air Strike Killed Al Qaeda Bomb Maker: Terror Big Also Trained 'Shoe Bomber' Moussaoui," ABC News, January 2006.

51 For a list of captured or killed al-Qaeda leaders, go to: www.msnbc.com/modules/wtc/wtc_globaldragnet/custody_alqaida.htm.

52 To illustrate the point, two physicians helped to execute a mid-2007 attack on the airport in Glasgow, Scotland. "NBC: U.K. Terror Arrests Include 2 Doctors; Officials Say Most, If Not All, of 5 People Now in Custody Are From MidEast." Available at: www.msnbc.msn.com/id/19522388/. More at length on al-Qaeda's recruitment

practices, Marc Sageman, *Understanding Terror Networks* (Philadelphia: University of Pennsylvania Press, 2004).

53 For the US proliferation assessments in this period of time, see the annual reports of the Arms Control and Disarmament Agency, entitled *Adherence to and Compliance with Arms Control Agreements*, and of the Department of Defense, which were entitled *Proliferation, Threat and Response*.

54 For more detail on the chemical weapons activities of these states, go to www.GlobalSecurity.org and to http://cns.miis.edu/research/cbw/possess.htm.

55 See William S. Cohen, *Annual Report to the President and the Congress* (Washington, DC: Office of the Secretary of Defense, January 1998), chapter 1.

56 For an analysis of the need to secure former Soviet chemical weapons and of efforts to keep former Soviet chemical weaponeers gainfully and peacefully employed, see Amy E. Smithson, "Improving the Security of Russia's Chemical Weapons Stockpile," in *Chemical Weapons Disarmament in Russia: Problems and Prospects*, report no. 17 (Washington, DC: Henry L. Stimson Center, October 1995), pp. 3–20; and *Toxic Archipelago: Preventing Proliferation from the Former Soviet Chemical and Biological Weapons Complexes*, report no. 32 (Washington, DC: Henry L. Stimson Center, 1999).

57 Liu Jianfei, "Contemplating the Threat of Biological Weapons Proliferation," in Amy E Smithson (ed.), *Beijing on Biohazards: Chinese Experts on Bioweapons Nonproliferation Issues* (Washington, DC: Center for Nonproliferation Studies, 2006), pp. 11–12. Available at: http://cns.miis.edu/pubs/week/pdf_support/070917_liu.pdf.

58 John Douglas and Mark Olshaker, *Unabomber: On the Trail of America's Most-Wanted Serial Killer* (Parsippany, NJ: Pocket, 1996).

59 Kubergovic threatened to bomb the US Congress with two tons of sarin. For a synopsis of the Kubergovic case, see Smithson, *Ataxia*, pp. 27–8.

60 For Mirzayanov's account of this weapons program and his experience, see "Dismantling the Soviet/Russian Chemical Weapons Complex: An Insider's View," in *Chemical Weapons Disarmament in Russia: Problems and Prospects* (Washington, D.C.: Henry L. Stimson Center, 1995), pp. 21–33.

61 The novel agents reported from the Czech Republic combine characteristics of VX and sarin to enhance lethality via both skin and inhalational exposure routes. J. Matousek and I. Masek, "On the Potential Supertoxic Lethal Organophosphorus Chemical Warfare Agents with Intermediate Volatility," *ASA Newsletter*, 94–5:1 (1994); J. Bajgar, J. Fusek and J. Vachak, "Treatment and Prophlaxis Against Nerve Agent Poisoning," *ASA Newsletter*, 94–4:10 (1994).

62 Note that the CWC does have a general purpose criterion that prohibits the use of any chemical for purposes of war. The *novichok* and Czech Republic agents are not listed on the CWC's Schedule 1, nor are all of the chemical components listed on the CWC's other Schedules of controlled chemicals.

63 Other than the physical size of a production facility that makes highly toxic chemicals, other telltale signs of this type of a facility that could fall by the wayside with the use of microtechnology are high stacks or heavy-duty ventilation and scrubbing equipment. International Union of Pure and Applied Chemistry, *Impact of Scientific Developments on the Chemical Weapons Convention*, Draft Report, Rev. 1, May 13, 2007.

64 Tuan H. Nguyen, "Security: Microchallenges of Chemical Weapons Proliferation," *Science* 309:5737 (2005): 1021.

65 Holger Löwe, Volker Hessel and Andreas Mueller, "Microreactors: Prospects Already Achieved and Possible Misuse," *Pure Applied Chemistry* 74 (2002): 2271–6; Ian Hoffman, "Scientist: Terrorists Could use Microreactors," *Oakland Tribune* (August 12, 2005).

66 For more on the advantages of microreactors, see International Union of Pure and Applied Chemistry, *Impact of Scientific Developments on the Chemical Weapons Convention*.

67 "Microreactors could redefine chemistry, nanodrip by drip," *Small Times* (December 8, 2003). Available at: www.smalltimes.com/articles/article_display.cfm?ARTICLE_ID=269148&p=109. Also, James E. Kloeppel "Ceramic Microreactors Developed for On-site Hydrogen Production," September 19, 2006. Available at: www.news.uiuc.edu/NEWS/06/0919kenis.html. University of Texas at Arlington, Public Affairs Office, "Microreactor Process Developed for Biodiesel Refining," press release, June 5, 2006; Clay Boswell, "Microreactors Gain Wider Use As Alternative to Batch Production," *Chemical Market Reporter* 266 (October 4, 2004): 8.

68 Michael Freemantle, "'Numbering Up' Small Reactors," *Chemical & Engineering News* 81 (2003): 36–7.

69 Nguyen, "Microchallenges of Chemical Weapons Proliferation," 1021; Löwe *et al.*, "Microreactors," 2272.

70 Robert D. McFadden, "India Disaster: Chronicle of Nightmare," *New York Times* (December 10, 1984); Stuart Diamond, "The Disaster in Bhopal: Lessons for the Future," *New York Times* (February 3, 1985); Jackson B. Browning, "Union Carbide: Disaster at Bhopal," in Jack Gottschalk (ed.) *Crisis Response: Inside Stories on Managing Under Siege* (Detroit, MI: Visible Ink Press, 1993).

71 The Australia Group's 40 member states have harmonized their export controls on high proliferation-risk chemicals, some dual-use chemical and biological equipment, and select human, plant and animal pathogens. For more on the Australia Group and its control lists, see: www.australiagroup.net. Members of the CWC are currently prohibited in trading the high-risk chemical precursors on Schedule 2 with states that have not joined the treaty. The CWC's Article VIII provides for adjustment of the treaty's prohibitions and controls to keep it abreast of technical advances. Moreover, the CWC mandates that Review Conferences "take into account any relevant scientific and technological development," and the CWC's Scientific Advisory Board is to discuss pertinent technical developments with an eye toward adjusting the treaty's provisions. On the CWC's provisions, governance and activities, see: www.opcw.org. The CWC's 2003 Review Conference paid scant attention to new chemical production processes. Nguyen, "Microchallenges of Chemical Weapons Proliferation," 1021. See also, Tuan H. Nguyen, "Implications of Advances in Production Technologies on the Chemical Weapons Convention," presentation given at IUPAC/OPCW Workshop on Impact of Scientific Developments on the Chemical Weapons Convention (Zagreb, April 23, 2007).

72 US Congress, Senate, Committee on Governmental Affairs, Subcommittee on International Security, Proliferation and Federal Services, *Hearing on Current and Future Weapons of Mass Destruction (WMD) Proliferation Threats*, 107th Cong., 1st Sess., November 7, 2001.

73 Nguyen, "Microchallenges of Chemical Weapons Proliferation," 1021.

74 For example, dedicated talks on microreactors have been given at the Annual Meeting of American Institute of Chemical Engineers and the International Symposia on Chemical Reaction Engineering. In 2006, the ninth International Conference on Microreaction Technology convened in Berlin. For the program, go to: http://events.dechema.de/events_media/Downloads/Geiling/Programm_IMRET_9_FINAL-view_image-1-called_by-events.pdf. A list of conferences sponsored by Micro Chemical & Thermal Systems through 2005 can be found at: www.pnl.gov/microcats/aboutus/conferences.html.

75 One of these companies increased its sales of microreactors by about 300 percent per year in the first three years of the twenty-first century. Marc Reisch, "Microreactors For the Chemical Masses," *Chemical & Engineering News* (December 1, 2004). Available at: http://pubs.acs.org/cen/news/8248/8248earlybus.html.

76 This agency defines a hazardous material as "any chemical that, if released into the environment, could be potentially harmful to the public's health or welfare," and lists 366 chemicals as hazardous. More information about US regulation of hazardous

materials can be found in the regulations of the Environmental Protection Agency, the Department of Transportation, or the Occupational Health and Safety Administration. Briefly, see Chris Hawley, *Hazardous Materials Response & Operations* (Albany, NY: Delmar Publishing, 2000), pp. 3, 200.

77 Safety is also the responsibility of chemical plant operators, who are also increasingly concerned with meeting process safety standards. If a plant operator mislabels tanks to mislead competitors, the local fire department will have the real codes for the tank contents in the event of an emergency. Marking of chemical storage tanks in Eastern Europe is not commonplace. Some chemical companies may require new employees to take a psychological profile test. Industry expert, PhD in chemistry, interview with author, Washington, DC (July 16, 2007).

78 Ibid. For more information on the problems of security at rail yards and ports, briefly see David Kocieniewski, "Despite 9/11 Effect, Railyards are Still Vulnerable," *New York Times* (March 27, 2006). Available at: www.nytimes.com/2006/03/27/nyregion/27secure.html?pagewanted=print. Also, *Risk Management: Further Refinements Needed to Assess Risks and Prioritize Protective Measures at Ports and Other Critical Infrastructure,* GAO-06–91 (Washington, DC: General Accounting Office, December 2006).

79 Smithson connected these dots in *Ataxia*, pp. 283–6. See also, Linda Greer, "New Strategies to Protect America: Securing Our Nation's Chemical Facilities," (Washington, DC: Center for American Progress, April 6, 2005).

80 These statistics, provided by the Environmental Protection Agency, stated that between 1,000 and 10,000 were in the injury/death zone surrounding 4,703 US chemical plants. *Homeland Security: Voluntary Initiatives Are Under Way at Chemical Facilities, but the Extent of Security Preparedness is Unknown*, GAO-03–439 (Washington, DC: General Accounting Office, March 2003), p. 10.

81 This figure constitutes about 10 percent of the roughly 5,700 chemical sites that the members of the CWC have declared as producing, processing or consuming above-threshold quantities of the treaty's controlled chemicals. Industry expert, PhD in chemistry, interview with author, Washington, DC (July 16, 2007).

82 On this attack series, Richard Weitz, "Chlorine as a Terrorist Weapon in Iraq," *WMD Insights* (May 2007). Available at: www.wmdinsights.com/I15/I15_ME1_Chlorine.htm.

5 Capacity-building and proliferation

Biological terrorism

Gabriele Kraatz-Wadsack

Introduction

Preventing bio attacks before they occur is obviously the most desirable goal – indeed, the adage of an "ounce of prevention" being worth a "pound of cure" is nowhere better illustrated than in the case of bioterrorism. The twenty-first century will be the century of biology and biotechnology. Advances in biology and biotechnology open new opportunities to benefit humanity but also the potential for the development and use of biological weapons (BW).

We have access to increasing technical ability and knowledge and unprecedented levels of information about the genetics of a variety of viruses and bacteria, the human body, life forms on this planet, etc. The use of biological sciences and technology to fight disease, hunger and pollution has enormous potential to help improving life and living conditions worldwide. Specifically, biotechnology has enormous potential to improve life for millions of people, using molecular diagnostics, recombinant drugs, new drugs and vaccine applications. Most obviously, biotechnology holds out the promise of the elimination of many types of infectious diseases. The speed and exponential development of biotechnology is comparable to the rapid developments which have taken place in the field of information technology. Unfortunately, some of these same technologies have the potential to be misused in order to produce biological weapons. The knowledge of how to improve production, stabilization and application of known and new substances is of dual-use nature and can serve for both peaceful as well as hostile purposes.

Without information technology, today's developments in genomics and proteomics would not have been possible. But information technology also provides huge data collections on diseases and pathogenic pathways that may uncover new vulnerabilities. Special challenges for the future are to improve global standards and increase international cooperation in addressing the challenge of biological risks. The lack of shared global language, risk assessment methodologies and standard settings in biosafety, biosecurity and best practices needs to be addressed. Therefore, global initiatives are needed to ensure that advances in the life sciences are used for the public good. Future national and international work must be focused on a global norm for biosafety levels, training curricula in safety and security, as well as multidisciplinary risk assessment.

All constituencies who share an interest in WMD non-proliferation have contributions to make in helping governments to maintain awareness of advances in the scientific domain. History shows that significant steps toward international peace and security can be accomplished through collective action.

While the 1899 Hague Convention and the Versailles Treaty of 1919 addressed the prohibition of the use of poisonous gases, the 1925 Geneva Protocol refers to chemical and biological weapons and prohibits the "use in war of asphyxiating, poisonous or other gases, and of Bacteriological Methods of warfare." These were early steps in the development of a fundamental international norm against the use of such weapons.

The first resolution adopted by the General Assembly on January 24, 1946[1] identified the goal of eliminating all weapons "adaptable to mass destruction." Numerous resolutions over the six decades to follow have echoed this goal, as elaborated in countless official statements by national delegations. The Secretary-General has promoted universal membership in the BWC and spoken repeatedly about the need for enhanced international cooperation against all types of terrorism, including biological terrorism. The UN remains, as stated in its Charter, "a centre for harmonizing the actions of nations" in achieving their "common ends" – it also has a specific mandate "to take effective collective measures for the prevention and removal of threats to the peace." Its universal membership also gives it an indispensable role in the debate and deliberation of global norms, including those relating to our subject today.

For example, at the time of the Geneva Protocol of 1925, chemical and biological weapons were linked together and their use was prohibited. Later, the chemical and biological weapons issues were only interlinked in the area of toxins. It seems that consideration was given to each of the categories posing their own and unique challenges, and they were considered individually rather than collectively. Disarmament efforts at the United Nations have for decades viewed the respective elimination of chemical and biological weapons as "partial measures" contributing to the agreed "ultimate goal" of general and complete disarmament, found in the preambles of both the 1972 Biological and Toxin Weapons Convention (BWC) and the 1993 Chemical Weapons Convention (CWC). The international community has chosen to pursue such partial measures on separate tracks.

Like the CWC, the BWC prohibits an entire class of weapons. Both conventions refer under each respective Article I to an all encompassing description of prohibitions. Under the CWC, the definition of chemical weapons is comprehensive and relates to all toxic chemicals and their precursors, except where intended for purposes not prohibited under the Convention, as long as the types and quantities are consistent with such purposes. Furthermore, a verification regime has been established to ensure that chemical activities are consistent with the objective and purpose of the treaty.

Under the BWC, the prohibition refers to the development, production, stockpiling or acquisition otherwise or retention of microbial or other biological agents, or toxins, whatever their origin or method of production, of types and in quantities that have no justification for prophylactic, protective or other peaceful purposes. Article I applies to all scientific and technological developments in the life sciences and in other fields of science relevant to the Convention. However, no verification regime exists under this treaty and no agency analogous to the OPCW exists, that is responsible for technical monitoring of compliance with the treaty.

States have implemented measures to regulate and control so-called dual-use technologies and materials to prevent their possible misuse. The measures consist of national legislation including criminal laws, physical protection of dangerous materials, export controls, disaster preparedness, international treaties and others. However, national implementation, as well as the application of laws and regulations, differ widely from state to state. After September 11, 2001, states have started to improve their legislation.

In 2004 the United Nations Security Council recognized the potential threat of the acquisition of nuclear, chemical and biological weapons and related materials, as well as their delivery systems, by non-state actors and mandated in its resolution 1540,[2] that states be legally obliged to provide for necessary legislation to counter such threats. Pursuant to the requirements of the resolution, states shall adopt and enforce appropriate effective laws that prohibit any non-state actor from developing, acquiring, manufacturing, possessing, transporting, transferring or using nuclear, chemical or biological weapons and their means of delivery, as well as prohibiting attempts to engage in, participate in as an accomplice, assist or finance any of these activities. Resolution 1540 (2004) is the first United Nations Security Council resolution that is directed to non-state actors' potential use of weapons of mass destruction (WMD).

Through its resolution 1540 (2004) the Security Council sent a strong message to prevent proliferation and placed requirements on states to prohibit the trade in nuclear, chemical and biological weapons; their related materials; and their means of delivery. Special emphasis was given to stopping support for such proliferation-related trade by non-state actors, including producers and manufacturers, financiers, logistical supporters and a range of individuals involved in the global supply chain. The resolution also includes trafficking as a form of proliferation threat to international peace and security, widening the potential non-state actor to include traffickers, brokers, technicians and scientists that have direct access to related materials. If 1540 is successfully implemented and enforced it will achieve major goals in non-proliferation: to make a difference by the end of the decade – if not before – and by representing not only one but a number of coordinated and innovative approaches to prevent the proliferation of nuclear, chemical and biological weapons, their means of delivery and related materials for terrorist purposes.

Other Security Council resolutions are in place which are related to counter-terrorism such as the Counterterrorism Committee and its Executive Directorate (Security Council resolution 1373[3]) and the Security Council Committee established pursuant to resolution 1267 (1999), also known as "the Al-Qaida and Taliban Sanctions Committee."

Another resolution to counter terrorism was adopted in September 2006 by the United Nations General Assembly, called the "Global Counter Terrorism Strategy".[4] This resolutions' adoption marks the first time that member states have agreed to a comprehensive and global strategic framework to counter terrorism. The strategy spells out concrete measures for member states to address the conditions conducive to the spread of terrorism, to prevent and combat terrorism and to strengthen their individual and collective capacity to do so, to protect human rights, and to uphold the rule of law while countering terrorism.

The practical steps to be undertaken include a wide array of measures ranging from strengthening state capacity to counter terrorist threats to better coordinating the United Nations System's counter-terrorism activities. The resolution's Annex contains a "Plan of Action" that addresses "measures to prevent and combat terrorism," under which – among other measures – the United Nations System was invited to develop, together with member states, a single comprehensive database on biological incidents, ensuring that it is complementary to the biocrimes database contemplated by the International Criminal Police Organization. The ultimate objective of the database is to provide information that will help planners and emergency responders – worldwide – to have a better understanding of the range of incidents and responses that have occurred in the past and to enable research and analysis to identify possible emerging trends and patterns, thus initiating and coordinating proactive activities to prevent and combat terrorism. Data generated within such a database will also be valuable in conducting analyses of responses to bioincidents and provide training for those in the security and response area. Data on lessons learned can provide better preparation for similar future events to detect, contain, treat and effectively manage a terrorist plot or event involving biological warfare agents.

In the context of the "Plan of Action," the member states also encouraged the Secretary-General to update the roster of experts and laboratories, as well as the technical guidelines and procedures, available to him for the timely and efficient investigation of alleged Biological Weapons (BW) use. The 6th Review Conference of the BTWC, which was held in November–December 2006, had noted that the SG mechanism, established in 1987, represents an international institutional mechanism for investigating cases of alleged BW or toxin weapons use and that the update of the technical guidelines was encouraged.

In addition to all of the above, the member states further noted the importance of the proposal of the Secretary-General to bring together, within the framework of the United Nations, the major biotechnology stakeholders, including industry, the scientific community, civil society and governments, into a common program aimed at ensuring that biotechnology advances are not used for terrorist or other criminal purposes, but for the public good, with due respect for the basic international norms on intellectual property rights.

The global market for materials and equipment, the global availability of requisite know-how, the fact that dangerous pathogens do not carry flags marking their national origin, the irrelevance of national boundaries for containing the spread of disease – all these are just some of the reasons why purely unilateral, national initiatives cannot alone suffice in meeting this challenge.

That is why the Secretary-General in 2006 suggested promotion of dialogue on a global level in order to sensitize decision-makers to non-proliferation issues with respect to biotechnology advances and the General Assembly noted the importance of this proposal.

The entry into force on July 7, 2007 of the International Convention for the Suppression of Acts of Nuclear Terrorism[5] is another example of international action to counter terrorist use of WMD.

All these international actions reflect and reinforce concerns about chemical, biological, radiological and nuclear weapons, their delivery means and related materials.

International organizations are also engaged in activities to counter the proliferation of dual-use goods and weapons-related materials. In addition to its safeguards activities, the International Atomic Energy Agency (IAEA) also undertakes to assist states in improving physical protection controls over nuclear materials, while the Organisation of the Prohibition of Chemical Weapons (OPCW) has the unique advantage of possessing an intrusive system of verification to confirm compliance with a fundamental norm of non-possession.

Ways and means to combat WMD proliferation

Addressing the geopolitical risk related to the proliferation of WMD and the measures to combat this proliferation includes intelligence, export controls, international arms control treaties and other mechanisms. Nodes of interaction are people, money, goods and the supply side. Some of the tools intended to reduce the risk of proliferation are called "indicators" and can be interpreted as "fingerprints" or "trail." A good indicator alerts to a problem before the problem is worse and helps to recognize what steps need to be undertaken. The prerequisite to understanding "indicators" is information and knowledge. Furthermore, generation of information, and the processing of it, creates new knowledge. This knowledge and assessment of indicators provide for a trend analysis which could lead to the detection of illicit activity, followed by interdiction and disruption.

Multilateral export control regimes have used indicators for interception and disruption of dual-use material in exports. Those regimes are consensus-based, voluntary arrangements of supplier countries that produce technologies useful in developing WMD and conventional weapons. The regimes aim to restrict trade in these technologies to keep them from proliferating to terrorist or states not part of these regimes.

The four principal regimes are the Australia Group[6] which focuses on trade in chemical and biological items; the Missile Technology Control Regime;[7] the Nuclear Suppliers Group;[8] the Wassenaar Arrangement[9] (previously CoCom),[10] which focuses on trade in conventional weapons and related items with both civilian and military (dual-use) applications. In addition to these regimes, there is a multilateral nuclear export control group called the Zangger Committee.[11]

Lists were developed within each of the regimes, which contain technical definitions of equipment, components, items, materials, software and technologies that are considered important for the development of illicit weapons.

Regime members conduct a number of activities, including (1) sharing of information about each others' export-licensing decisions, including certain export denials and, in some cases, approvals; and (2) adopting common export control practices and control lists of sensitive equipment and technology into national laws or regulations.

Since September 11, 2001, all export control regimes have acted to address terrorism. For example, the Australia Group added counter-terrorism as an official purpose of the regime and added a number of items to its control list in an effort to control the types of items that terrorists, rather than states, would seek in order develop chemical or biological weapons. These items included toxins, biological equipment and the transfer of information and knowledge that could be used for chemical and biological weapons purposes. In addition, in June 2002 the Australia Group adopted a provision in its new guidelines for licensing sensitive chemical and biological items that made it the only regime to require its members to adopt "catch-all" controls – so-called end-use controls. "Catch-all" controls authorize a government to require an export license for items that are not on control lists but that could contribute to a WMD proliferation program if exported. However, export control regimes are not universal and the degree to which norms in trading are being observed varies widely around the world.

In addition to transfer, export controls and supply-side restrictions for some activities or materials, there is a need for all countries and competent institutions to provide bioweapons awareness training for life scientists working in the public and private sectors. This provides also for "societal citizen verification" which can be understood as whistle-blowing.

All these international actions reflect and reinforce concerns about chemical, biological, radiological and nuclear weapons, their delivery means and related materials.

No single solution to the overall challenge exists. Multilateral arms control conventions are a part of the significant, global response to effectively counter chemical and biological weapons proliferation. The risk of the possible misuse of science and technology must be minimized through the engagement of the entire international community, but without hampering life-science research. For this reason, awareness of the risk factors of life-science research has increased substantially over the past several years. Legal, scientific, security, public health and law enforcement experts from around the world have worked together to explore ways to improve coordination and cooperation to increase security, without limiting the research necessary for the development of medication and vaccines. They have explored how know-how can be protected from being exploited for malicious purposes and how to help build a culture of awareness, responsibility and accountability among the scientific community.

As emphasized by the international Weapons of Mass Destruction Commission, chaired by Hans Blix,[12] such weapons pose significant and real threats to international peace and security. Terrorists look at weaknesses of states and explore possibilities to access materials for illicit purposes. Efforts to redress this through treaties and national implementation measures provide the most effective antidote against such exploitation. Only universal adherence to international/multilateral disarmament and non-proliferation agreements and full and effective implementation of their provisions by the states parties can provide assurance of an effective prohibition of biological and chemical weapons, and contribute to the non-proliferation of WMD.

The scope and role of procurement in general

The underlying task for assessing indicators is to assess intent. This intent relates to acquisition of a capability and to actors. However, since the actors cannot easily be identified, awareness needs to be strengthened to recognize signals that attract further attention and follow-up. In the export control sense, it is "know your customer." If a baseline has been established, a change analysis for deviation from this established baseline of the customer can serve as indicator.

Since proliferation may have indigenous and external procurement dimensions, the acquisition of a capability plays a major role. Acquisition of a capability can be based upon knowledge on what to acquire, informal and formal ties to procurement entities, financial assets or smuggling. The related activities may depend on the import of foreign technology, arms, equipment, tools, parts and materials or on the reverse engineering or otherwise indigenous acquisition.

An acquisition network performs logistics, financial and administrative functions, but it may not be directly involved in the technical effort. Movements of part of material, immaterial and financial flows may appear uncontrollable to the extent that flows do not involve the recourse to legally established companies.

Procurement in the field of biology and biotechnology

The acquisition of a biological capability such as a fully assembled weapon may be considered difficult. Access to and trafficking of some of the related dual-use materials and technologies to produce biological weapons may be considered less difficult, depending on the degree of assets and know-how available, choice of candidate BW agents, scale of program, etc. Scenarios could be the acquisition of classic biological warfare agents with state assistance, working from scratch, engineering novel agents with advanced technologies, etc.

The biological area in its entirety has no choke points or "telltale" signatures or early detection of whether activities are for peaceful or other uses.

Illicit biological activities can be hidden easily in institutions that have legitimate interests in microbiology, genetics and fermentation. Because of the dual-use nature, there are many check points which would include export-control measures, border controls or others.

Many technologies in the biological area, including the advanced technologies such as biotechnology, have "dual-use" applications, offering simultaneous beneficial and malevolent uses. There are no unique signatures for fixed or mobile sites for BW agent research, development, production, filling or storage. In the case of a surveillance program or monitoring of activities, the net of check points would be huge and would need to cover all civilian activities, such as biological equipment production capacity (e.g. for fermenters, incubators, dryers), biological laboratories (e.g. in hospitals, universities, public health, food-testing institutions), biological production facilities (e.g. vaccine production, drug formulation and production), food industry (e.g. dairy facilities, breweries, distilleries), agricultural facilities (e.g. pesticide operations, herbicide spraying, greenhouses), import and supply agencies, storage facilities, etc. It also covers the dual-use nature of military defense programs.

Regarding equipment, no high-tech needs to be involved. Even biological containment is not required for the production of the BW agents. BW agents can be produced in very small quantities, e.g. in five-liter flasks. Even with this small-scale production capability, the aggregate production provides material for weaponization.

The main factors influencing indicators are related and depend upon the intended activity and its scale, e.g. weaponization, storage considerations, etc. The scale of the activity influences the acquisition sources and the indicators for early warning regarding acquisition by imports or otherwise. Other factors are the degree of possible utilization of country resources such as government and private laboratories, the knowledge available, the adaptability, flexibility, the practicality of the intended activity and the means of meeting accepted risks. Indicators will vary, depending if activities are centralized, compartmentalized or dispersed to different locations. The judgment about the necessity of accepting certain risks and the choice of the candidate agent will influence whether arrangements for biosafety will be made. Any indicator to uncover covert and/or compartmentalized activities is related to the individual proliferation areas, which are: recruitment of know-how; acquisition of materials and facilities; the supply of a turnkey facility with the logistics thereof; and the financial flow or transfer, etc. Thus, essential functions necessary in a supplier's network can be identified as providing a product or a specific component as well as management of production with know-how and knowledge.

Biological equipment and supplies are critical to establish a biological warfare capability. Besides equipment, the technical effort would require microbial agents, or biologics. The biotechnology sector encompasses pharmaceuticals, medicine, agriculture, biomaterials and defense applications. Microprocess technologies with future desktop molecular and pharmaceutical manufacturing and the increase in "made-to-order genes" significantly impact on the control of chemical and biological materials. Future trends in the pharmaceutical area include "designer drugs" – those designed specifically for the patient. Drug delivery, in combination with medical device technology, is an increasing field for easier application and different routes for application of medicine, with new designs of integrated and combined material systems that can solve specific problems – for instance, drug delivery to particular sites in the body. In the medical area, immunization technology by aerosol delivery has increased gradually. Modern products work faster and more effectively, and simultaneous application to more than one target at the same time is possible.

In the agricultural sector the new approaches play a significant role for food safety and food security. Benefits for the agricultural sector are through the enhanced detection and monitoring of pathogens in crops, food and grains, as well as through control measures and resistance against disease in the field. The production of artificial seeds by mapping genomes of important crops, leading to a uniformity of crops, is also possible.

New possibilities of microbial degradation and bioremediation can help to clean up polluted soil and water.

Modern genetic techniques provide ways to engineer novel pathogens to introduce properties that enhance their utility as offensive weapons or to embrace manipulations of key body functions. Newly introduced properties may include altered host range, enhanced infectivity and person-to-person spread, increased difficulty of detection, ability to circumvent available treatment or protective vaccination, improved survival of the pathogens in diverse environments and greater resistance to inactivation by decontamination agents. This may also include new knowledge. The knowledge of how to improve production, stabilization and application of known and new substances is of dual-use nature and can serve for both peaceful as well as hostile purposes.

Procurement in the case of Iraq's BW program and lessons learned

The lessons learned from Iraq's BW program are relevant in many aspects. The entire BW program derived from the state intelligence and security apparatus. The first institute for the acquisition of a BW capability was created to conduct scientific, academic and applied research in the field of chemistry, physics and microorganisms. The BW program was very well guarded, secret and compartmentalized. Very little information transpired internationally or triggered concern before 1991 when Iraq invaded Kuwait, although the large imports could have raised some concerns before.

Another lesson learned is that the working conditions for the scientists were without biological containment since this was considered unnecessary for the production of biological warfare agents by Iraq. There was no high-tech equipment and some of the BW agents were produced in five-liter glass bottles and the amounts produced (over 2,000 liters) were weaponized.

In the end, when the international community was alerted, the offensive BW program was exposed by the international community consisting of governments who were providing export information and the international inspectors who could compare the exported materials with the actual materials for a material balance account. If items and materials were consumed it was most likely that this was for the prohibited program and not for civilian purposes.

If there was a need, civilian facilities and equipment were taken over for the production of BW agents.

The lessons learned from Iraq's BW program illustrate the problems and the importance of procurement information. With respect to Iraq's weapons programs, the United Nations Special Commission (UNSCOM) established that during the period from the mid-1970s to 1990, foreign suppliers had provided major critical technology, equipment, items and materials. The suppliers included government agencies and organizations, private companies, together with individuals who acted as brokers and middlemen. Some 80 branches of foreign banks outside of Iraq were involved in acquisition transactions. In addition, dozens of trans-shipment companies were involved in the delivery of items and materials to Iraq. While there were cases when suppliers were aware of the final use of the equipment and materials delivered to Iraq, there were also many cases when the providers were unaware of the intended end-use of the items they sold to Iraq.

When trade restrictions came into effect, Iraq experienced increasing problems in importing technology and raw materials from states that had implemented appropriate licensing systems. Iraq then switched its procurement efforts to companies or subsidiaries that operated in countries where such measures had not yet been developed, introduced or fully implemented. It also involved its own organizations which stood for legitimate commercial activities as a front importing company. Thus, some laboratory equipment and materials used by the BW program were procured through the Ministry of Agriculture, Ministry of Oil and Ministry of Health, and machines and tools for missile projects through the Ministry of Industry. Further adjustments included the use of networks of brokers and middlemen, with offices registered in third countries. Contracts and end-user certificates could thus be issued to Iraq's front importation companies and foreign trading companies, instead of directly to Iraq. Foreign trading companies then, acting on behalf of brokers and middlemen, would procure the required goods from manufacturers and distributors. To further cover the final destination of goods, if required, brokers and middlemen would arrange for multiple trans-shipments by freight handlers. The goods would not be delivered to Iraq but to a neighboring country in the region, where they would be transported to Iraq by an Iraqi shipping company acting on behalf of the end-users or their agencies. The length of the "procurement chain" depended on the geographic location of the manufacturers, the sensitivity of the item and existing trade regulations in their countries. Accordingly, the creation of additional bank accounts in multiple foreign banks was required to support such a sophisticated procurement mechanism at each phase and location of its function. This resulted in proportionally increased costs for the items and materials procured in this manner. Mindful of the difficulties it had experienced in the acquisition of dual-use equipment and materials, and the likelihood that such difficulties would increase in the future, Iraq procured some items in excessive quantities in order to secure and meet possible needs in the future.

In the area of biological weapons, procurement of equipment used for BW research and development and production of BW agents, bacterial isolates and complex growth media, was from foreign suppliers. The paper trail of the growth media import led to the uncovering of the true purpose of this excessive import. Some 42 metric tons were supplied to Iraq. However, no credible explanations were provided for this importation. The growth media was said to have been imported on behalf of the Ministry of Health for the purposes of diagnostic laboratories in hospitals. This importation of media by types, quantities and packaging was assessed by UNSCOM as grossly out of proportion to Iraq's stated requirements for hospital use. Iraq explained the excessive quantities imported and the inappropriate size of the packaging as being a "one-of-a-kind mistake" in order to justify the import for medical diagnostic purposes. However, for hospital diagnostic purposes, only 200 kg per annum were declared as consumed prior to this.

A further incongruity was that, out of all the types of media required for diagnostic purposes, only a select few were imported. Most of the growth media import was suitable for the production of anthrax and botulinum toxin, rather than for diagnostic purposes. Furthermore, the packaging was another indicator. Diagnostic assays use very small quantities of media and so, because the media deteriorates rapidly once a package has been opened, media for diagnostic purposes is normally distributed in 0.1–1 kg packages. The manufacturer's guarantee for quality assurances was for 4–5 years, which would make the excessive quantity imported unusable for hospital diagnostics. However, the media imported by Iraq in 1988 was packaged in 25–100 kg drums. This style of packaging is consistent with the large-scale usage of media associated with the production of biological agents such as in vaccine production.

Growth media was never on any list of an export control regime and thus is below the radar. There is no export license required and no end-use control. Some growth media are more "dual-use" than others; for example, yeast extract is a nutrient and exists in many food products in solid or liquid form, but it can also be used in anthrax production. Therefore, it would be important to have information about "special orders," which would be supplied in excess in all aspects.

The equipment used in Iraq's BW production program was largely taken from biological facilities in Iraq that had earlier acquired the equipment for legitimate purposes. These were, for example, animal vaccine production equipment which was dismantled, transferred and used for the BW program.

Given the critical role that dual-use technology, equipment and materials acquired from foreign suppliers played in Iraq's development of its WMD programs, the evaluation of procurement data proved to be one of the major tools to uncover Iraq's secret BW activities. The procurement data were audited and analyzed. The "audit trail" was a combination of information, documents and records relating to specific actions undertaken by Iraq for the acquisition of items and materials. This included communications and negotiations with prospective suppliers, tenders describing services required, relevant specifications of items, offers made by suppliers, preparation and implementation of contracts, insurance documents, bills of landing, trans-shipment information, customs documentation and final delivery certifications by end-users.

Documentation also included a number of financial statements such as the opening of operational accounts in corresponding banks, issuing letters of credit and a variety of money transfers from the accounts of end-users in Iraq to the banks involved in the transactions.

This experience of UN verification in Iraq shows that despite extensive concealment policies and practices, it was still possible to find evidence (an audit trail) of procurement activity. The nature and scope of the procurement process was such that multiple "fingerprints" of past acquisitions existed not only at various organizations in Iraq, including ministries and agencies, establishments and banks, but also outside Iraq, in countries of suppliers and third-party countries through which goods were trans-shipped.

Although the introduction of export licensing by individual States significantly slowed down and limited Iraq's procurement efforts prior to 1991, it did not stop them completely. The provisions involving licensing of the exports on the grounds of only end-user certificates without on-site verification were not able to fully solve the problem of possible trans-shipments of dual-use items and materials to Iraq. Iraq's demonstrated ability to make adjustments and modifications to its procurement techniques to overcome trade restrictions, and to a certain degree even under sanctions, shows that only a combination of national measures to control the export of dual-use items and materials with a universal international mechanism for export/import notifications and on-site verifications can provide a sufficient degree of confidence that these goods are not used for proscribed purposes.

The mechanism for export/import monitoring under Security Council resolution 1051 (1996) is an example of the efficient functioning of trade regulations. It displays, however, that when one of its complimentary elements, on-site inspections, was not in force due to the absence of the inspectors in Iraq from 1999 to 2002, its effectiveness was diminished.[13]

The need for global norms

A network of mutually reinforcing provisions and innovative approaches are needed – National Committees for example – which would establish a central authority over many departments that already deal with transportation of dual-use materials. Global norms are needed to deal with a global threat.

The task of preventing bioterrorism is clearly one of the most challenging tasks facing all countries – yet it is also a burden that no single country can bear entirely alone. The global market for materials and equipment, the global availability of requisite know-how, the fact that dangerous pathogens do not carry flags marking their national origin, the irrelevance of national boundaries for containing the spread of disease – all these are just some of the reasons why purely unilateral, national initiatives cannot alone suffice in meeting this challenge. Such initiatives, while certainly necessary, are far from sufficient.

Though the BWC has no verification regime, it remains as relevant today as it was in 1972 and offers a positive case study of how international law adapts to changing times. The ideas to address the diverse challenges of the twenty-first century are multiple. It is noteworthy that, in the context of the inter-sessional meetings prior to the 6th Review Conference of the BWC in 2006, the issue about codes of conduct for scientists was raised.

The states parties to the BWC have again taken effective action and have agreed to continue an inter-sessional process in the years ahead until the next Review Conference in 2011, and to address the following subjects: national implementation, including law enforcement as well as regional and sub-regional cooperation on BWC implementation; improving biosafety and biosecurity measures as well as oversight, education and awareness raising, for example, through codes of conduct for scientists to prevent the misuse of bioscience and biotechnology research; and on enhanced international cooperation, assistance, and exchange in biological sciences and technology for peaceful purposes, including disease surveillance.

The BWC review process offers one appropriate forum for promoting international cooperation. It may in time play a more important role as its states parties come to recognize their common need for this treaty to have greater institutional support. The launching of the Implementation Support Unit to the Convention is one concrete step forward in the right direction, and other steps may follow.

Yet the BWC, despite its great value in creating a global taboo against the possession or production of biological weapons, has not yet achieved universal membership. The actual and potential threats of biological terrorism, however, are indeed fully global in scope and it only seems appropriate that the United Nations would have some important role of its own to play in reinforcing the fundamental norms embodied in that treaty.

Just as its member states alone cannot eliminate the risk of bioterrorism, nor can the organization of the United Nations – this task will require initiatives at all levels of international society, including national policies, bilateral and plurilateral agreements, regional arrangements and fully multilateral undertakings. In conclusion, the prohibition, non-proliferation, prevention of terrorist use and disarmament of unconventional weapons requires a multidisciplinary effort. Science and technology play a vital role in this effort for security and protection and to keep verification measures up-to-date. A web of interaction is needed and must involve networks of scientists, policy-makers, industry, civil society and the public at large.

At best these efforts can only substantially reduce – but never entirely eliminate – this risk, yet that is itself a worthy goal.

Notes

1 UN General Assembly resolution 1 (1), 1946.
2 S/RES/1540 (2004).
3 S/RES/1373 (2001).
4 A/60/288 (2006).
5 The Convention was adopted by the General Assembly in 2005.

6 The Australia Group was established in 1985.
7 The Missile Technology Control Regime was established in 1987.
8 The Nuclear Suppliers Group was established in 1975.
9 The Wassenaar Arrangement was established in 1996.
10 Coordinating Committee for Multilateral Export Controls, established in 1947.
11 The Zangger Committee was formed in 1971.
12 WMD Commission, *Weapons of Terror, Freeing the World of Nuclear, Biological and Chemical Arms* (Stockholm, 2006).
13 United Nations Monitoring, Verification and Inspection Commission, *UNMOVIC Compendium June 2007.*

6 Terrorism and potential biological warfare agents

Walter Biederbick

It is difficult to find a good definition of a biological warfare (BW) agent or a biological weapon. In the Bacteriological (Biological) and Toxin Weapons Convention (BWC) a very general definition is given. It reads:

> 1. Microbial or other biological agents, or toxins whatever their origin or method of production, of types and in quantities that have no justification for prophylactic, protective or other peaceful purposes. 2. Weapons, equipment or means of delivery designed to use such agents or toxins for hostile purposes or in armed conflict.

For the purpose of this study BW agents can be defined as pathogens, such as bacteria, fungi, viruses or toxins of biological origin that can be intentionally used in an act of warfare to cause injury or death to humans. Currently, about a dozen agents are considered high-priority pathogens for the development of medical biological warfare countermeasures. These agents are also known in this content as the "Dirty Dozen." However, there are many more microorganisms and toxins that have the potential to be employed as BW agents. The Australia Group lists 70 human pathogens, the Directorate General SANCO of the EU-Commission evaluated about 174 potential agents and the United States Centers for Disease Control and Prevention named 23 agents as pathogens with the potential for weaponization or use for terror purposes.

The focus here is limited to human pathogens. Animal and plant pathogens are not covered by this chapter. However, enormous economic damage can be created by the misuse of these pathogens by terrorist groups. Natural outbreaks of foot and mouth disease in cloven-hoofed livestock or avian influenza in poultry show that billions of euros worth of assets can easily be destroyed by the introduction of these agents into an agricultural area.

It has to be clear that state-supported programs and activities of non-state actors differ in the intent and some other critical aspects. State-supported programs do not have to fear the prosecution of law enforcement agencies, they have better access to scientific knowledge, better logistic support and usually more financial resources. However, the aims of state-supported military programs are in general more challenging than these of non-state actors. Biological weapons are, like nuclear and chemical weapons, "political weapons." They provoke a maximal retaliation and their use is banned either by international agreements or by the "international public opinion." So the use of these weapons should give the attacker such a significant strategic advantage that the conflict is solved in his favor and the payoff is more than the damage caused by the response to the use of a "political weapon." To achieve this strategic edge the biological weapons have to be highly effective and reliable. This requires a sound scientific approach with a large research and development program, the development of an integrated system that includes not only the BW agents but also the delivery system, testing and storage of the weapons, operational military planning for the use and training of the personnel. Terrorists do not have to meet all these challenges, because their main goal is to create fear and expose the inability of the targeted government to cope with the situation.

The range of scenarios for the use of potential BW agents varies from assassinations such as we have seen in the case of Georgi Markov in 1978, to limited attacks like the 1984 incident created by members of the Bhagwan Shree Rajneesh group or the *Bacillus anthracis*-containing letters seen in 2001 on the east coast of the United States, to doomsday scenarios with casualties well in the 100,000s or more. These doomsday scenarios are mainly based on studies and field trials carried out in the United States and the United Kingdom after World War II until BW was internationally banned in 1972. Two studies can be quoted as open accessible references: the report of the Secretary General of the United Nations from 1969; or the World Health Organization from 1970.[1] However, it is difficult today to evaluate the scientific reliability of reports of this type, because the original data from scientific experiments and field trials are not easily accessible.

This chapter will focus in detail on the BW agents listed as Category A by the United States Centers for Disease Control and Prevention because most publications and studies dealing with BW are limited on these agents. Nevertheless, some misleading assumptions on potential BW agents have been introduced into the published literature and have created some misunderstandings in the assessment of potential BW agents. Therefore it seems worthwhile to focus again on them in order to give a different opinion on some critical points. However, it is well understood that a complete review of all potential BW agents should not be limited to the Category A ones. A complete analysis is, however, a major undertaking and will go beyond the scope of this chapter.

For the purpose of this chapter the potential BW agents listed in Category A by the United States Centers for Disease Control and Prevention are divided into three groups:

- man-to-man transmissible or contagious microorganisms:
    - smallpox virus
    - *Yersina pestis* (plague)
    - viral hemorrhagic fever viruses like the Marburg virus, Ebola virus, Lassa virus and Crimean-Congo fever virus

- non-man-to-man transmissible microorganisms:
  - *Bacillus anthracis* (anthrax)
  - *Francisella tularensis* (tulareamia, dear fly fever)
- toxins:
  - *Clostridium botulinum* toxins (botulism)
  - ricin.[2]

Man-to-man transmissible microorganisms

Some pathogens with the potential to be used as BW agents have the special characteristics of an infection and can be spread directly from person to person, creating secondary and tertiary waves of infected victims. These secondary victims may not have been in the area of the initial attack, and these persons can become ill even weeks or months after the attack. This aspect makes BW agents unique compared to chemical warfare agents or nuclear weapons, which means these BW agents can create additional fear because there are no safe areas left and the normal countermeasures of the security agencies are not sufficient without additional anti-epidemic public health measures.

Fortunately, only a few of the agents in the "Dirty Dozen" are transmissible from person to person. These agents are the smallpox virus, *Yersinia pestis* (*Y. pestis*) as the causative agent for plague, and viral hemorrhagic fever viruses like Marburg virus or the Ebola virus.

The *smallpox virus* was a worldwide natural threat to mankind over thousands of years. The World Health Organization (WHO) has declared it eradicated since 1980.[3] There are only two declared repository stocks of the smallpox virus worldwide: i.e. the Institute VECTOR in Koltsovo, Russia and the Centers for Disease Control and Prevention in Atlanta, USA. It can be assumed that it would be extremely difficult and unrealistic for non-state-supported actors to access the smallpox virus in these locations. Whether there are other places in the world where the smallpox virus is kept secretly can only be speculated. Currently there is no sufficient proof of clandestine stocks of the virus in other locations, neither from the scientific perspective nor from intelligence evidence. Therefore, for the purposes of this study, the smallpox virus is considered inaccessible to non-state actors. It was reported that the smallpox virus was part of the Soviet biological weapons program before 1992.

The disease can be spread from person to person. Data from India in the 1960s, and an analysis of the smallpox outbreaks in Germany after 1945 show the requirement of a contact closer than 2 m to the infected patient or direct contact to linen or clothing used by the infected person. However, there are two anecdotal reports of smallpox transmission through the air over distances greater than 2 m.

Currently there is no approved treatment of smallpox. Animal experiments show some effects from antivirals like the investigational new drug ST-246 or the antiviral Cidofovir.[4] However, Cidofovir is know to have significant side-effects and is not considered to be suitable for general use.

Vaccination is the only countermeasure that has shown to be effective.[5] Therefore many countries such as the United States, United Kingdom or Germany have stockpiled sufficient amounts of vaccine for their populations. However, severe side effects of the smallpox vaccine are more frequent than in modern vaccines against other diseases. Mass vaccination would be an option to control the situation in an effective way only in case of a major smallpox outbreak.

Yesinia pestis causes the disease known as plague. Plague is endemic in several parts of the world. Like other endemic bacteria of viruses, it is theoretically possible to acquire *Y. pestis* from natural sources. However, this access would require time, money and specialized knowledge to find and select the appropriate strain to be used as a BW agent. *Y. pestis* was part of the BW programs of several governments in the twentieth century. Japan used plague-infected fleas to attack northern parts of the Chinese front before and during World War II. The effects are not well documented. Some sources report negligible success, while others report more than 100,000 victims from these attacks.

The pulmonary form of plague can be easily transmitted from person to person within a short distance (under 2 m). Therefore, this pathogen is considered to be capable of creating secondary outbreaks. The victims of secondary outbreaks could present in places very distant from the scene of the attack, and this could occur days after the attack. Due to the fact that the disease can by prevented and treated by today's antibiotics, sufficient medical countermeasures are available to limit the spread of a plague outbreak.

Y. pestis is sensitive to ultraviolet (UV) waves and dies rapidly in sunlight. This is one major limitation in the use of *Y. pestis* as a BW agent, and it is the reason the Japanese used fleas, rather than other dispersion devices, to transmit the bacteria and protect it from sunlight. Even more modern BW programs have not been able to overcome the challenge of *Y. pestis*' UV-sensitivity, thus it would also be very difficult for a terror group to overcome this challenge in the near future.

Viral hemorrhagic fever viruses like *Marburg* or, especially, *Ebola* are very popular when it comes to fictional scenarios in popular literature or movies. Most outbreaks in sub-Saharan Africa generate patient figures in the order of some dozens to a few hundred, and benefit from the poor hygiene standards and cultural habits (for example, preparation of bush meat and morgue rituals) in the affected areas. They can be controlled using basic hygiene measures. The transmission of the disease usually requires a very close contact to infected materials such as blood. Special skills are required to cultivate the viruses. There are little data available in the open literature about the stability of these viruses in the environment. From the epidemiology and the data generated under laboratory conditions it can be concluded that both viruses are much more sensitive to environmental influences than *Bacillus anthracis* or *Francisella tularensis*.[6] Therefore, the viruses have to be stabilized in order to use them as an aerosol in a biological attack. Additional technical skills, scientific knowledge and experience are required to achieve this.

There are often media reports stating that the Aum Shinrikyo cult tried to acquire the Ebola virus during an outbreak in Congo. However, it was not possible to confirm this information during the official investigation.

A high lethality rate of 90 percent has been observed during some outbreaks of Ebola fever or Lass fever. There is neither a specific treatment nor a vaccine available for these viruses. This combination of deficits creates fear and is a good basis for fiction.

Medical countermeasures against Marburg fever virus and Ebola fever virus infections are limited to isolation methods,[7] barrier nursing[8] and symptomatic treatment. Currently there are several research projects under way – mainly in the United States – to develop vaccines and other treatment options against Marburg virus and Ebola virus. Licensed vaccines or therapeutic antibodies are not yet available however.

Lassa virus and *Crimean-Congo fever virus* are other viral hemorrhagic fever viruses that are man-to-man transmissible. Lassa fever is endemic in western Africa and it is estimated that between 100,000 and 300,000 instances of the disease occur each year in the endemic area. Only 1–5 percent of the infected persons experience the hemorrhagic fever.[9] Most infections (80 percent) do not produce clinical symptoms.[10] The Lassa virus is spread much more easily than the Ebola or Marburg virus and it is considered to be much more accessible in the endemic areas. The major disadvantage of the Lassa virus as a biological agent is the low "attack rate," i.e. only a limited percentage of the infected persons develop a severe disease. This makes it impossible to plan or predict the outcome or success of an attack with the Lassa virus. The Crimean-Congo fever is currently endemic in the Balkans, in countries around the Black Sea and around the Persian Gulf. Ticks are the main route of transmission from animals to humans, but man-to-man transmission has been reported, especially in hospitals. Direct contact with infected blood, urine or other bodily fluids is considered as the main reason for man-to-man transmission. Infectious aerosols are also regarded as a possibility for man-to-man transmission. Depending on the subtype of the Crimean-Congo hemorrhagic fever virus the lethality varies from less than 1 percent up to 50 percent in patients admitted to hospital.

Ribavirin is recommended as a treatment option both for Lassa fever and Crimean-Congo hemorrhagic fever. The availability, the side-effects and the price limit the use of the drug to a few individual cases. A post-exposure prophylaxis with Ribavirin in larger populations is currently not possible due to the reasons given above. Stempidine (developed for the treatment of HIV-1 and HIV-2 infections) showed promising results in experimental animal Lassa virus infections. Future studies will have to prove the efficacy of this drug in human cases of Lassa fever virus infections.

Non-man-to-man transmissible microorganisms

Bacillus anthracis (*B. anthracis*) is the causative agent of the disease anthrax. Anthrax is a zoonosis that is endemic in most countries in the world. Therefore, theoretically the bacteria can be considered readily available. According to media reports, the Iraqi government acquired the anthrax strain for their BW program from the American Type Culture Collection (ATCC), a non-profit organization that collects and sells cell cultures, tissue, bacteria, etc. This saved the Iraqi government years of development in selecting a highly pathogenic strain from domestic outbreaks of anthrax. It can be assumed that terrorist groups today would have much more difficulty accessing highly pathogenic *B. anthracis* strains either in nature or through institutional culture and strain collections.

B. anthracis has been the "workhorse" in BW programs throughout the twentieth century. The imperial German BW program was directed against the horse supply of the allied armies in World War I and used a suspension of *B. anthracis*. In World War II, the British forces undertook major efforts to set up a strategic BW capability against German cattle herds. It was reported that in the second half of the twentieth century *B. anthracis* was a part of the British, United States, USSR and Iraq BW programs. The Aum Shinrikyo cult not only used botulinum toxins and sarin as terror weapons, but also disseminated *B. anthracis* from the roof of a building in Tokyo in 1993. Fortunately the terror group only gained access to a harmless strain of *B. anthracis* that is normally used to vaccinate cattle. Therefore no damage to humans was reported and the attempted attack remained undetected for years. In autumn 2001, several letters containing weapons-grade spores of *B. anthracis* were sent to people in several cities on the east coast of the United States. During this attack, 22 people contracted anthrax, five of whom died. This very limited attack illustrated the potential of *B. anthracis* as a BW agent. According to media reports and speculations from non-governmental organizations, it is believed that this attack was caused by a perpetrator who had access to weapons-grade anthrax spores under the control of the US government. If this speculation is correct, the anthrax letters of 2001 provide additional evidence that it is not easy to produce weapons-grade *B. anthracis* spores without access to the resources of a state program.

B. anthracis is very stable in the environment. Under the right conditions the spores can survive for years and remain viable. These characteristics, combined with the extremely high lethality of the pulmonary form of anthrax, make *B. anthracis* highly attractive for military programs.

Man-to-man transmission of anthrax is very unlikely. In 2001 an infant contracted the cutaneous form of anthrax only by exposure to contaminated clothing.

There are effective antibiotics available against *B. anthracis* infections if the treatment starts early enough (within 24–48 hours) and no antibiotic resistance has been introduced.[11,12] The treatment and the post-exposure prophylaxis should be continued for 60–100 days. If a large group of persons has been exposed to aerosolized *B. anthracis* it could a logistical problem to supply enough antibiotics. As an alternative, a combination of antibiotic treatment and vaccination is recommended.[13]

Francisella tularensis (*F. tularensis*) can cause a disease called dear fly fever, rabbit fever, or tularemia. There are different sub-types of *F. tularensis* distributed across the world. The more lethal sub-type, *T. tularensis*, is found mainly in North

America[14] (it has a case fatality rate up to 40 percent or higher, depending on the manifestation of the disease). The strains predominantly found in Europe and Russia cause a milder form of the disease.[15] Therefore, theoretically the lethal strains can be considered available in some parts of the world. However, it can be assumed that terrorist groups would have at least the same difficulties accessing highly pathogenic *F. tularensis* strains as they would have for *B. anthracis*. Additionally, it should be mentioned that the cultivation of *F. tularensis* is a challenge that requires a lot of experience and special media for the cultivation. These growth media are not well documented in the open literature. Therefore, accessibility and cultivation of *F. tularensis* can be a major hurdle for terrorist groups.

F. tularensis was part of the BW program of the United States in the 1950s and 1960s. There are no confirmed reports of the use of *F. tularensis* as a biological weapon.

F. tularensis is considerably stable in the environment over days and weeks, and only a few germs are required to cause the disease if inhaled. These characteristics make *F. tularensis* attractive for military purposes. However, orders of magnitude more germs are necessary after an oral ingestion in order to cause an infection.

Man-to-man transmission of tularemia is extremely rare and an absolute exception. The disease can be transmitted via vectors such as ticks.

There are effective antibiotics available against *F. tularensis* infections. The treatment should be continued for 10–14 days. The logistic burden of an antibiotic resupply is smaller compared to the situation after a *B. anthracis* attack. A licensed vaccine is not available in most European countries.

Toxins

Clostridium botulinum toxins are produced by different bacteria strains called *Clostridium botulinum, Clostridium baratii* or *Clostridium butyricum*. There are seven different types of botulinum toxins known, four of which are pathogenic in humans. Botulinum toxin A is the most toxic substance know to mankind. The bacteria are available in nature and cause natural food-borne disease outbreaks worldwide, particularly during the preservation or storage of food where the deadly toxin is reproduced under exclusion from oxygen. The number of botulinum intoxications in Germany decreased over the last decades to about a dozen per year (see Figure 6.1).

Sources have been found on the internet that describe the cultivation and extraction process in enough detail that botulinum toxins could be reproduced by a mid-level, non-specialized biologist. It is currently believed that there are still hurdles for a terror group that aims to produce botulinum toxins, particularly the difficult-to-access strain of *Clostridium botulinum* that easily produces the toxin. According to several reports, the Aum Shinrikyo cult in Japan attempted, unsuccessfully, to use botulinum toxins as a BW agent in Japan before the sarin nerve agent (chemical warfare agent) attacks in Matsumoto (1994) and Tokyo (1995). There is no information available as to why the attacks failed.

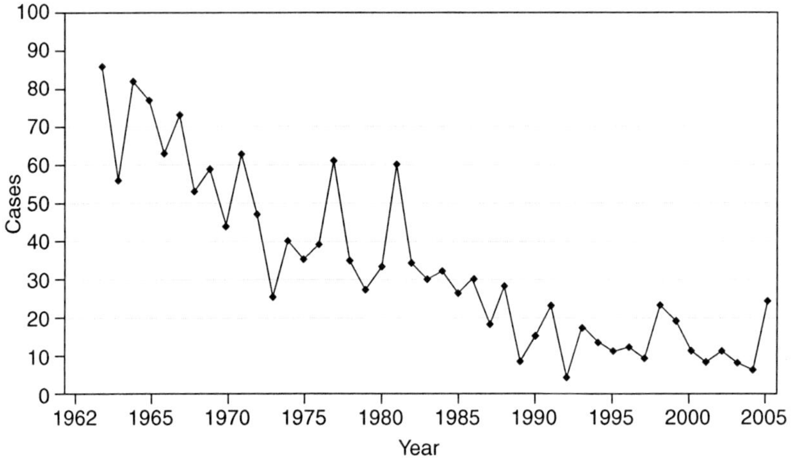

Figure 6.1 Botulism in Germany, 1962–2005.

Botulinum toxins are not very stable in the environment. The decay rate of aerolized botulinum toxin A is reported to be between 1–4 percent per minute. Botulinum toxins are much more stable if they are in a protein-containing liquid with a moderately acidic pH value.

A specific anti-toxin against botulinum toxins is available in many countries. The anti-toxin has a major limitation. It can only be effective if given early. After the symptoms have developed completely the anti-toxin cannot counteract the circulating toxin. The support of the vital functions is crucial for the survival of the patient.

Ricin is the only potential BW agent in this study not listed as Category A by the United States Centers for Disease Control and Prevention. There is one main criterion for the selection of BW agents, and that is the accessibility of the agent. There is a very wide range of accessibility among the "Dirty Dozen." For example, ricin is one of the easiest BW agents to obtain, while the smallpox virus is one of the most difficult. This is the reason ricin deserves a special focus.

The castor bean, which is the natural source of ricin,[16] can be bought in any well-equipped garden supply store in Europe, or can be grown in sufficient quantities in a backyard in central Europe. Furthermore, the extraction process of ricin from the castor bean can be carried out even with low-quality technical equipment with sufficient efficacy. Therefore, ricin is considered to be readily available. There have been several media reports over the last decade about successful or attempted criminal homicides in the United States using ricin, as well as reports on suspected attempts by United Kingdom terror groups in 2003 to extract ricin, and the appearance of ricin samples in the United States mail system in late 2003. As a matter of fact, in the 2003 United Kingdom case, the available evidence was not sufficient for prosecution.

Ricin was part of the BW program of the United States in the 1940s and of Iraq until 1991. The assassination of Mr. Markov in 1978 in London is frequently quoted. There are several reports from the second half of the 1990s on the use of ricin for criminal homicides.

Ricin is orders of magnitude less toxic than other toxins.[17] Therefore it is not the agent of choice for a large-scale attack.

There are no specific medical countermeasures available for the treatment of ricin intoxications.[18] Vaccines[19] and therapeutic antibodies are under development.

Threat assessment versus risk assessment

There are different ways to look at the problem of assessing the probability and other specific factors connected to bioterrorism or BW. Mainly, two approaches are used in parallel or separately:

1. The observation of activities by means of intelligence agencies and/or other security/law enforcement agencies. In old terms the word "threat" reflected this point.
2. The assessment of vulnerabilities based on the characteristics of the potential BW agent and the population targeted. Here the words "risk assessment" would be the right term.

It seems to be clear that neither a single threat assessment nor a risk analysis alone could give a complete picture of the situation. Therefore, currently a combination of both ways seems to be an optimized way to approach the problem.

It is very true that the past may not be a good predictor for non-state actors' views on the use of potential BW agents. This is true not only for terrorist groups, but also for state-supported programs. However, leaving this way has at least two risks. On one hand there is the risk of losing the way and ending up completely in speculation. On the other hand, if done well, it can not be excluded that this process leads to a "recipe book" for bioterrorism. So, a meta-analysis of the few incidents that happened already can provide some insights in the deficit in the algorithm of an analysis process. A list of incidents or BW programs is proposed, which seems to be appropriate for this meta-analysis (see Table 6.1).

By looking to the incidents or program it seems obvious that five of the seven have been disclosed by a defector or a criminal investigation on other business. One was identified after a large-scale military operation as a "by-product." So it seems to be very easy to hide even a large scale BW program. No science-based indicators, neither generic nor specific, could be identified. Even worse, the absence of a program could not be proven prior to a massive military operation and an unprecedented extensive countrywide search for BW agents.

A general problem in assessment is the quality and reliability of the information available. As an example, the "weapons grade" quality of the *B. anthracis* spores in the letters from 2001 can be used. In the media it was reported that the material from the letters contained four times more *B. anthracis* spores than the

Table 6.1 A list of incidents and BW programs

Soviet program	Until 1992
Iraq program	Until 1991
Iraq (alleged – not existing) program	From 1998 to 2003
Al-Qaeda/Afghanistan	Until 2002
Bhagwan Shree Rajneesh group	1984
Amu Shinrikyō cult	1993
B. anthracis-containing letters	2001

best material from the Soviet program prior to 1992. A Russian expert in *B. anthracis* research commented on the quality of the spores in the letters that "such poor material would not have passed the quality control process in the times before 1992."

"Information warfare" comes as an additional complication in the analysis process. Even the assessment of intelligence agencies is often based to a large degree on open-source information. The intended distribution of misleading information can surely influence the assessment process in a way that cannot be underestimated. An example for this is given in the report of the Commission on the Intelligence Capabilities of the United States regarding Weapons of Mass Destruction (2005).

The majority of the studies on BW countermeasures are carried out in the United States. Only nuclear or biological weapons remain as a strategic threat, if someone takes into account the specific aspects of the strategic situation of the continental United States. With conventional weapons, even the combined forces of the rest of the world would not represent an existential challenge. Strategic chemical warfare requires massive logistic support, easily comparable to conventional warfare. The existential nuclear threat of the Cold War was reduced significantly in the last 15 years and will remain exclusive to the very few powers who can afford a strategic nuclear program. So the use of BW agents seems to be the Achilles' heel for the United States.

The strategic European security situation is different. In the Balkans we have had two major military operations within the last 15 years. The distinction between this and the situation of the United States is that in Europe we directly neighbor several areas of instability and insecurity. Therefore, conventional threats are much closer to Europe than to the United States and different priorities for strategic threats can easily be explained.

History

It has already been mentioned that there have been attempts by terror groups or criminals to employ BW agents. Between 1990 and 1995, the Aum Shinrikyo cult attempted to use *B. anthracis* and botulinum toxins on several occasions. These events were disclosed in the criminal investigations following the sarin attack in 1995. In 1984, members of the Bhagwan Shree Rajneesh cult poisoned several salad buffet bars in the town of The Dallas in Oregon with *Salmonella typhimurium* in order to influence the local elections. This incident was reported to the law enforcement agencies by a person leaving the cult. In 2001 the anthrax letters on the east coast of the United States infected 22 people and killed five, creating a worldwide fear of bioterror and a massive demand for countermeasures against emerging bioterror threats.

Future

The foreseeable technical development in life sciences and genetic engineering will lead to new opportunities to create new and advanced BW agents. It can be anticipated that this knowledge will proliferate and that the technical capability to create BW agents will be available to more regions in the world and to groups and individuals in countries where governmental control is limited. This will increase the risk of BW attacks with new or modified agents.

It is very expensive to develop and maintain a nuclear capability. It is said that the equivalent of hundreds of millions US dollars are required to achieve this. It may be questionable whether it is possible to create a doomsday scenario with 100,000 victims or more in a single attack with chemical warfare agents. A strategic chemical capability would require quantities in an industrial scale, as well as the appropriate means of delivery and logistics. The financial resources to achieve this quickly pass the $10 million level. The financial support required to generate a capability for either nuclear or chemical mass destruction goes most probably beyond the potential of most terror groups. The cost for a basic, non-sophisticated program using potential BW agents may be orders of magnitude lower. Biotechnology is becoming more and more available and cheaper and more affordable to terrorist groups.

Dissemination/employment of potential biological warfare agents

In most cases, only a very small amount of a BW pathogen is needed to harm a human. An often-quoted calculation of the WHO from 1970 states that 50 kg of weapons-grade *B. anthracis* spores released in an aerosol attack will kill 125,000 people in a city of 500,000 inhabitants. To achieve this with any known chemical warfare agent is not technically possible.

Compared to chemical warfare agents or nuclear weapons, BW agents require relatively little logistical support for production. However, the capability to disseminate BW agents as an aerosol requires profound knowledge not only in microbiology, but also in aerosol techniques. This limits the threat of BW agents because only state-run programs or groups supported by states seem be capable of mastering these technical challenges up to now. Not even Saddam Hussein's regime in Iraq shared its knowledge in this field with terror groups.

In the media it is reported from time to time that terrorists consider combining improvised explosive devices with BW agents as a method of dissemination.

This makes little sense. Most BW agents are very sensitive to heat and would not survive the high temperatures of an explosion.

Therefore, it seems more likely that an attack using BW agents would focus on the contamination of food and water. Compared to civilian societies, military forces in an area of operations normally have a tight control system for food and water supplies. This limits the risk of a large-scale attack. If these systems fail, there can be severe consequences. This was demonstrated in the recent operations in Afghanistan, where NATO forces suffered hundreds of temporary casualties due to self-limiting gastro-intestinal infections. The infections were a consequence of insufficient inspection of food supplies bought from the local/regional markets, which were contaminated with pathogens endemic to those communities.

Because of the heat sensitivity of most BW agents, an attack on a food supply would be more successful if it was carried out at the end of the preparation process, just before consumption. A potent-toxin poisoning of a dessert buffet in the dining facility of an exposed office building could create hundreds of casualties, and this could be completed by terror groups with a limited technical background.

Fortunately, large-scale water supplies for cities or regions are much more difficult to contaminate. Because BW agents lose most of their efficacy from dilution in water or the purification process, water supplies would have to be poisoned with very large amounts of a pathogen to cause negative health effects. Large-scale attacks on a water supply network of a city or a region seem too challenging, even for very capable groups. However, assaults on a water pipe system of a house, a street or a small area could be much easier to achieve.

Notes

1 World Health Organization (1970) *Health Aspects of Chemical and Biological Weapons* (Geneva: WHO), pp. 98–109.

2 US CDC Category B agent.

3 Fenner, F., Henderson, D.A., Arita, I., Jezek, Z. and Ladnyi, I. (1988) *Smallpox and its Eradication* (Geneva: WHO).

4 Bray, M., Martinez, M., Smee, D.F., Kefauver, D., Thompson, E. and Huggins, J.W. (2000) "Cidofovir protects mice against lethal aerosol or intranasal cowpox virus challenge," *Journal of Infectious Diseases* 181 (1): 10–19.

5 Henderson, D.A., Inglesby, T.V., Barlett, J.G., Ascher, M.A., Eitzen, E., Jahrling, P.B., Hauer, J., Layton, M., McDade, J., Osterholm, M.T., O'Toole, T., Parker, G., Perl, T., Russell, P.K. and Tonat, K. (1999) "Smallpox as a biological weapon," *Journal of the American Medical Association* 281 (22): 2127–37.

6 Mitchell, S.W. and McCormick, J.B. (1984) "Physicochemical inactivation of Lassa, Ebola, and Marburg viruses and effect on clinical laboratory analyses," *Journal of Clinical Microbiology* 20: 486–9.

7 Fock, R., Koch, U., Wirtz, A., Peters, M., Ruf, B. and Grünewald, T. (2001) "First medical and anti-epidemic measures if a viral haemorrhagic fever is suspected (Erste medizinische und antiepidemische Mabnahmen bei Verdacht auf virales hämorrhagisches Fieber)" *Medizinische Welt* 52: 126–32.

8 Esbroeck, van M., Groen, J., Hall, W., Heyman, P., Niedrig, M., Tegnell, A., Vaheri, A., Vandenfelde, C. and Zeller, H. (2001) "Management and control of Viral Haemorrhagic Fevers and other highly contagious viral pathogens," *Scientific Advisory*

Committee, 2nd version (Berlin: European Network for Diagnostics of Imported Viral Diseases (ENIVD)). Available at: www.enivd.de/NETZ.PDF, pp. 1–48.

9 Schmitz, H., Köhler, B., Laue, T., Drosten, C., Veldkamp, P.J., Günther, S., Emmerich, P., Geisen, H.P., Fleischer, K., Beersma, M.F., Hoerauf, A. (2002) "Monitoring of clinical and laboratory data in two cases of imported Lassa fever," *Microbes and Infection* 4: 43–50.

10 Richmond, J.K. and Baglole, D.J. (2003) "Lassa fever: epidemiology, clinical features, and social consequences," *BMJ* 327: 1271–5.

11 Biederbick, W., Fock, R., Güttler, K. and Veit, C. (2002) "Infektionen durch Bacillus anthracis," *Deutsche Medizinische Wochenschrift* 127: 809–14.

12 Jones, M.E., Goguen, J., Critchley, I.A., Draghi, D.C., Karlowsky, J.A., Sahm, D.F., Porschen, R., Patra, G. and DelVecchio, V.G. (2003) "Antibiotic susceptibility of isolates of Bacillus anthracis, a bacterial pathogen with the potential to be used in biowarfare," *Clinical* Microbiology and *Infection* 9 (9): 984–6.

13 Fowler, R.A., Sanders, G.D., Bravata, D.M., Nouri, B., Gastwirth, J.M., Peterson, D., Broker, A.G., Garber, A.M. and Owens, D.K. (2005) "Cost-effectiveness of defending against bioterrorism: a comparison of vaccination and antibiotic prophylaxis against anthrax," *Annals of Internal Medicine* 142 (8): 601–10.

14 Petersen, J.M., Schriefer, M.E., Carter, L.G., Zhou, Y., Sealy, T., Bawiec, D., Yockey, B., Ulrich, S., Zeidner, N.S., Avashia, S., Kool, J.L., Buck, J., Lindley, C., Celeda, L., Monteneiri, J.A., Gage, K.L. and Chu, M.C. (2004) "Laboratory analysis of tularemia in wild-trapped, commercially traded prairie dogs, Texas, 2002," *Emerging Infectious Diseases* 10 (3): 419–25.

15 Reintjes, R., Dedushaj, I., Gjini, A., Jorgensen, T.R., Cotter, B., Lieftucht, A., D'Ancona, F., Dennis, D.T., Kosoy, M.A., Mulliqi-Osmani, G., Grunow, R., Kalaveshi, A., Gashi, L. and Humolli, I. (2002) "Tularemia outbreak investigation in Kosovo: case control and environmental studies," *Emerging Infectious Diseases* 8 (1): 69–73.

16 Arena, J.M. (1981) "Plants that poison," *Emergency Medicine* 13: 25–57.

17 Knight, B. (1979) "Ricin – a potent homicidal poison," *BMJ* 1 (6159): 350–1.

18 Challoner, K.R., McCarron, M.M. (1990) "Castor bean intoxication," *Annals of Emergency Medicine* 19 (10): 1177–83.

19 Medina-Bolivar, F., Wright, R., Funk, V., Sentz, D., Barroso, L., Wilkins, T.D., Petri, W., Cramer, C.L. (2003) "A non-toxic lectin for antigen delivery of plant-based mucosal vaccines," *Vaccine* 21 (9–10): 997–1005.

7 Influence diagram analysis of nuclear and radiological terrorism

Charles D. Ferguson

Despite some terrorist groups having expressed interest in nuclear and radiological terrorism, terrorists have yet to detonate a nuclear weapon or disperse radiation using a radiological weapon. What, if anything, is stopping terrorists from carrying out such attacks? Are they self-deterred? Are they repulsed by the idea of these types of terrorism? For certain political–religious and apocalyptic terrorists, the answer is definitely no. For example, Usama bin Laden has clearly stated that it is al-Qaeda's duty to acquire weapons of mass destruction. Also, when Shoko Asahara was in charge of the apocalyptic group Aum Shinrikyo, he expressed strong interest in obtaining nuclear weapons, and Aum demonstrated its ability to make chemical weapons. But other terrorist groups with strong ties to constituencies or national territories could be self-deterred from nuclear or radiological terrorism. For instance, national-separatists would tend to avoid contaminating their homeland with radioactivity or run the risk of massive retaliation against their supporters in that territory. Chechen rebels, a national-separatist group trying to free Chechnya from Russia, however, have at least demonstrated the capability of launching a radiological attack. For example, in 1995 Chechen rebels placed radioactive cesium in Ismailovsky Park in Moscow and called a television crew to the scene. They did not, however, disperse the radioactive material.

Are technical impediments holding back terrorists from making nuclear and radiological weapons? These weapons require greater skill than improvised explosive devices, although security experts concur that a crude radiological weapon is within the capabilities of most terrorist groups as long as they can acquire the radioactive material. In a related question, are terrorists having difficulty in finding and hiring technically competent people? Are financial barriers too high, especially in buying nuclear explosive material? Are the risks of being scammed by hucksters or ensnared in sting operations too great for terrorists to seriously pursue these types of terrorism? These are some of the questions that need to be addressed when considering whether terrorists will converge onto nuclear and radiological weapons.

Nuclear and radiological terrorism are complex multidimensional issues involving multiple actors. There are no quick and easy answers as to whether or when terrorists will launch nuclear or radiological attacks. Faced with this situation, an analytic tool is needed that can wrestle with the complexity. One such tool is decision analysis. It offers a methodology for facilitating clear thinking about complicated problems and for formulating a logical framework for qualitatively and quantitatively assessing various solutions.[1] Decision analysis can help determine the best course of action through the thicket of entangled problems. It provides a structured framework for organizing information about the problem, gathering additional information if necessary, evaluating the information, and deciding on a clear course of action. Importantly, this analytic framework also takes into account uncertainties associated with the state of knowledge about the issue and factors in the risk tolerance of the decision-maker. Here, the decision-maker is the leader of a terrorist group or someone in the group who has the power to marshal resources and launch an operation. First, let's examine the analytical factors that this decision-maker would have to consider when deciding whether to launch a nuclear or radiological attack. This examination will provide a set of identifiable early-warning mechanisms for nuclear and radiological terrorism. To model the role of the authorities working to counter terrorism, the examination can also include a second decision-maker. The two decision-makers can be thought of as two agents in a non-cooperative game in which each tries to achieve opposing goals while making moves to thwart the other.

Nuclear terrorism fundamentals

Before beginning the decision analysis of nuclear and radiological terrorism, it is necessary to define terms. Here, nuclear terrorism refers to two types of acts: (1) acquiring an intact nuclear weapon from a nation's arsenal and detonating it; or (2) stealing or buying nuclear explosive material and making an improvised nuclear device, which is a crude nuclear bomb, and then detonating it. Experts agree that nuclear terrorism is a very low probability but very high consequence event. Based on what is known about terrorist capabilities and the technologies of making nuclear weapons, there are some reasonable assumptions to factor into a study of nuclear terrorism.

The first assumption is that a terrorist group cannot make a sophisticated nuclear weapon such as a thermonuclear or hydrogen bomb. This type of nuclear weapon requires the technical skills and industrial capacity only available to some nation-states.

Second, terrorist groups with enough technical competence have a non-zero probability of building crude nuclear explosives given access to sufficient nuclear explosive material. Crude nuclear explosives could make use of the two first-generation nuclear weapon designs. The most basic design is the gun-type bomb in which, like a gun, the bomb shoots a lump (bullet) of highly enriched uranium (HEU) down a gun barrel into another lump of HEU, a kind of nuclear explosive material. Each lump is shaped into a sub-critical piece of nuclear explosive material. When these two lumps combine, they form a supercritical mass of nuclear explosive material, leading to a detonation. The actual yield of the explosion would depend on the design details of the bomb. As long as the two sub-critical pieces of HEU combined, there is a good chance that an appreciable yield would result. For example, the Hiroshima bomb used a gun-type design and produced about 13 kilotons of TNT-equivalent yield. This bomb destroyed the heart of a city and killed about 100,000 people soon after the explosion.

The other first-generation design is called an implosion device and is considerably more sophisticated than the gun-type bomb. The implosion bomb squeezes, or implodes, HEU or plutonium into a very dense state to form the supercritical mass. If the squeezing does not proceed smoothly, the device could become a dud or at best create a "fizzle," resulting in an explosive yield much below the design yield. Thus, implosion bombs require well-shaped explosive lenses to squeeze the nuclear material and very fast electronic triggers to simultaneously ignite the many components of the explosive lens. In sum, a reasonably competent terrorist group has a much better chance of making a gun-type bomb than an implosion-type bomb. But the gun-type bomb can only use HEU to produce a relatively high yield. If plutonium were used, it would assuredly result in a fizzle gun bomb, but this could still serve the terrorists' purposes if they wanted to disperse plutonium and cause a very low-yield nuclear explosion.[2]

Third, terrorist groups cannot make nuclear explosive material, which includes HEU and plutonium.[3] HEU is defined as a mixture of uranium isotopes containing 20 percent or greater proportion of uranium-235. Uranium-235 easily fissions if its nuclei absorb slow- or fast-moving neutrons. However, natural uranium only consists of 0.725 percent uranium-235, which is too low of a concentration to sustain an explosive chain reaction. Uranium enrichment is the process used to increase the concentration of uranium-235. Low-enriched uranium, typically containing 3–5 percent uranium-235, is suitable for fueling many electrical power-producing reactors but cannot fuel nuclear bombs. Nuclear bombs require much greater concentrations of uranium-235, usually 90 percent, which is considered weapons-grade, although conceivably any type of HEU could fuel a nuclear weapon. Typically, concentrations of 80 percent or more uranium-235 are considered weapons-usable. Terrorists do not have the technical and industrial resources required to enrich their own uranium. Perhaps in the future, technological advances in laser enrichment could offer technically sophisticated terrorist groups the capability of enriching their own uranium, but that possibility appears remote. Thus, they would have to seize HEU from existing stockpiles created by nation-states.

Similarly, producing plutonium is outside the capabilities of terrorist groups without significant assistance from a nation-state. Plutonium-239, the fissile isotope that can fuel nuclear bombs has a relatively short half-life of 24,000 years and thus, on the geological time scale, it decays rapidly. Therefore, it must be made artificially. Nuclear reactors can produce plutonium. A plutonium-producing reactor can cost tens to hundreds of million dollars and require many technicians to build and operate it. Extracting the plutonium from spent nuclear fuel requires a reprocessing facility and the capability of safely handling highly radioactive fission products in the spent fuel. Consequently, terrorists are highly unlikely to make their own plutonium and would have to seize it from existing stockpiles created by nation-states.

How much highly enriched uranium and plutonium are there and where are these nuclear explosive materials located? HEU stockpiles are relatively abundant. There are about 1,850 metric tons available globally – sufficient fissile material to make tens of thousands of nuclear bombs.[4] Militaries control most of the HEU. In the military sector, Russia and the United States possess approximately 1,700 metric tons of HEU for weapons purposes and naval propulsion. China, France and the United Kingdom control tens of metric tons of HEU. Pakistan has an estimated HEU stockpile of several hundred kilograms for its weapons program. In the civilian sector, more than 40 countries have HEU. While the HEU holdings at particular sites are often too small to fuel a nuclear bomb, many of the more than 120 research reactors and related facilities hold enough HEU in each site to make a nuclear weapon.[5] Security experts tend to agree that the HEU stockpiles most vulnerable to theft are those located in Russia, Pakistan and several countries with civilian nuclear facilities.

Like HEU, plutonium exists in civilian and military sectors. Military stockpiles hold more than 250 metric tons of plutonium – enough to make tens of thousands of nuclear bombs. As with HEU, the United States and Russia possess most of the military plutonium, more than 90 percent. They have declared about 100 metric tons as excess to defense needs but have yet to render this material into non-weapons-usable forms. China, France, India, Israel, North Korea, Pakistan and the United Kingdom hold the remainder of the military plutonium. The United Kingdom has declared 4.4 tons as excess to its defense needs. France, Russia, the United Kingdom and the United States have stopped producing plutonium for weapons purposes, and China may also have stopped military plutonium production. In contrast, India, Israel and Pakistan plan to continue to produce plutonium for their weapons stockpiles. In July 2007 North Korea shut down its plutonium-production facility at Yongbyon as part of the February 2007 agreement at the Six Party Talks.

Like military plutonium, civilian plutonium poses a nuclear weapons risk. Reactor-grade plutonium in the civilian sector can fuel nuclear weapons as long as the plutonium has been separated from highly radioactive fission products in spent nuclear fuel.[6] (These fission products provide a protection barrier against theft or diversion of the plutonium embedded in the spent fuel.) Because several countries continue to separate or reprocess plutonium from spent fuel and because the rate of consumption of this plutonium lags behind the rate of production, the civilian stockpiles continue to grow. The rate of growth has been roughly ten metric tons per year in recent years. This yearly amount could fuel hundreds of nuclear bombs. Presently, more than a dozen countries possess more than 230 metric tons of separated plutonium. This global civilian stockpile is comparable to the global military plutonium stockpile.

There are concerns that bulk handling facilities for plutonium could pose vulnerabilities for insider diversion of plutonium.[7] Despite international safeguards applied to these facilities, several hundred kilograms of plutonium have not been accounted for. While it is believed that this unaccounted for material is trapped in piping and other components of these facilities, a business-as-usual approach to the presence of a large amount of material unaccounted for could be exploited by insiders. They could reason that over a long enough period of time they could divert enough plutonium for a bomb as long as the separate instances of diversion are within the error bars of the material unaccounted for that is normally expected at the plutonium-processing facility. This insider scenario is far from easy to carry out, but it does point out possible shortcomings in the ability to adequately account for and safeguard nuclear explosive material when several metric tons worth are handled at large facilities.

Radiological terrorism fundamentals

Next, let's turn to radiological terrorism to define basic terms and discuss the radiological assets that terrorists could target. Like nuclear terrorism, radiological terrorism consists of two different activities: (1) obtaining radioactive material and dispersing it or using some other mechanism to release radiation; and (2) attacking or sabotaging nuclear facilities such as nuclear power plants, plutonium reprocessing plants and spent fuel storage pools, to release radiation. Unlike nuclear terrorism, radiological terrorism cannot produce a nuclear explosion. Radiological terrorism would likely harm relatively few people immediately after the release of radiation, although over a period of several years to decades hundreds to thousands could suffer health effects due to radiation exposure depending on the radiation release scenario. Therefore, radiological terrorism is not considered a weapon of mass destruction but has been described as a weapon of mass disruption.[8]

Concerning the first type of radiological terrorism, a radiological weapon could use different mechanisms to release radiation. For instance, a so-called dirty bomb would employ conventional explosives to disperse radioactive material. A dirty bomb is a particular type of radiological weapon more generally known as a radiological dispersal device (RDD). An RDD need not use conventional explosives depending on the chemical composition or physical state of the radioactive material. For example, radioactive material in a powdered form such as cesium chloride could be dispersed by wind or dissolved in water. In contrast to an RDD, a radiation emission device (RED) would not disperse the radioactive material but instead would emanate radiation from stationary radioactive material. An RED could be an effective terror weapon in crowded locations such as major train stations. Although an x-ray machine could act as an RED by emitting radiation, it would need an electrical power supply for it to produce and emit radiation. In contrast, an RED using a radioactive source would not need a power supply – it is always "on" and emitting radiation. Finally, a radiological incendiary device (RID) would combine radioactive material with incendiary material in order to employ fire to spread radioactive contamination.[9]

Despite the different release or dispersal mechanisms, all radiological weapons need radioactive material. This material can be found in numerous applications. For instance, spent nuclear fuel contains highly radioactive fission products. Globally, there are tens of thousands of tons of spent nuclear fuel. Without special handling equipment for spent fuel, terrorists would risk exposing themselves to a lethal dose of radiation within minutes. In other applications, millions of radioactive sources are used throughout the world in industry, medicine and scientific research. However, only a small fraction of these sources are considered as a high security risk.[10] Nonetheless, that fraction still contains thousands of radioactive sources that could cause significant damage if used in a radiological weapon. Like spent fuel, some of the commercially used sources are highly radioactive and thus would require special handling procedures to avoid receiving a lethal dose from an unshielded source in a short period of time. But there are still thousands of sources that have intermediate levels of radioactivity and could pose attractive targets for acquisition by terrorists.

Determining the security risk of a radioactive source requires knowing how much radioactivity it contains, whether the radioactive material is in a readily dispersible form (for use in an RDD), whether the source is portable, and what security measures are in place to guard the source. Focusing only on the radioactivity content, sources with large amounts of radioactivity would pose a greater security concern than those with lesser amounts. However, as indicated earlier, terrorists might be dissuaded from handling highly radioactive sources that could result in a lethal dose within a short period of exposure. Even suicidal terrorists need to live long enough to deliver a radiological weapon. For example, terrorists would likely not consider seizing very highly radioactive sources from food and sterilization irradiation plants, which contain millions of curies (tens of thousands of terabecquerels) of radioactivity. But sources containing upwards of hundreds or thousands of curies (tens to hundreds of terabecquerels) could pose a risk of terrorist use.[11] These sources can be found in teletherapy machines for treating cancer, blood and research irradiators, and radiography cameras for checking pipe welds and related industrial applications. Such sources could cause significant harm if dispersed in an RDD and could be handled by terrorists who had some knowledge of radiation safety practices.

Where radioactive sources are used significantly impinges on the security risk. For instance, oil exploration sites use oil well logging sources, which are very portable and have been stolen on several occasions. Also, hospitals and universities can pose an increased risk for theft of radioactive sources because, by design, these facilities are accessible to the public and tend to possess relatively powerful radioactive sources for medical applications and scientific research. Pertinent questions for determining the security risk are: What is the amount and quality of security training? What physical security measures are used? What are the security procedures and how thoroughly are they carried out?

In recent years, adequacy of the physical security at nuclear power plants, research reactor facilities and other nuclear facilities has attracted increased attention from governments and non-governmental analysts. The nuclear industry in the United States and some other countries has been proud to point out that it has rigorous security at its nuclear power plants, although some independent analysts disagree with this assessment. So far, terrorists have not launched a full-scale attack at such a facility. The target environment is relatively rich. In dozens of countries there are about 440 commercial nuclear reactors in operation and a roughly comparable number of research reactor facilities. Spent fuel pools, especially at commercial power plants, contain huge amounts of radioactivity, which could cause significant contamination if released to the environment. But causing a radioactivity release from these pools or the reactors is very challenging. Almost all of the power plants, for example, have defense-in-depth safety and security features including a containment building as a last line of defense to prevent the release of a significant amount of radiation to the environment, although in recent years there has been a renewed concern about the ability of the containment structures to withstand the impact of a large, high-speed airliner.

Influence diagram approach

Let's turn to applying a tool of decision analysis to try to understand the decisions terrorists would have to make in determining whether to launch a nuclear or radiological attack. The tool is known as an influence diagram, described as: "[A] graphical tool used to capture the essence of a problem and facilitate communication among multi-disciplined teams and the decision board."[12] In the following applications, the multi-disciplined teams are various terrorist cells that have certain skills and abilities. The decision board or decision-maker is the terrorist leadership who has the authority to order the use of the terrorist organization's resources to carry out operations.

Do terrorist groups operate this way? Do they consider various proposals and then determine the best or an effective proposal that meets their objectives? A detailed answer lies beyond the scope of this analysis. However, it is worth pointing to evidence that al-Qaeda operates in this fashion for at least its complex and high-profile operations. For example, the 9/11 Commission noted that al-Qaeda's leadership considered airplane crashes into a nuclear power plant as well as other complicated airplane attacks before deciding on the 9/11 operation.[13] Also, Dhiren Barot, an al-Qaeda operative, had developed different plots and wrote a report to al-Qaeda leadership arguing for the efficacy of these plots. One of the plots would have involved using radioactive material from 100 or more smoke detectors to make a dirty bomb.[14] (Millions of smoke detectors would be needed to acquire sufficient radioactive americium to make a potent dirty bomb.) Barot was arrested in August 2004 and was sentenced in May 2007 to 30 years in jail. Unlike the 9/11 operatives, he had not carried out any of his planned attacks.

Like a business decision, a decision to commit an act of terrorism involves evaluation of limited or uncertain information, considering the risk of failure versus success. A bad business decision could result in failure or bankruptcy of the business. A bad terrorism decision could result in arrest or other type of interdiction of the individual terrorist or terrorist cell. Even if a terrorist group does not formally use decision analysis, effective leadership of the group requires at least an evaluation of uncertainties, risks of failure and alternative courses of action.

To understand the fundamental components of an influence diagram, let's first examine an influence diagram involving some of the major components needed to decide whether to launch an attack using an intact nuclear weapon. In influence diagrams, decisions are represented by rectangles; uncertainties or chance variables are represented by ovals; and objective variables (quantities that the decision-maker is trying to maximize or minimize) are represented by hexagons. An arrow represents an influence. If X influences Y, knowing X would affect the expectation about the value of Y. However, an influence does not necessarily mean a causal relation between X and Y, nor does it necessarily imply a flow of something such as data, material or money. This observation underscores that influence diagrams are not flow charts and are not cyclical. An influence diagram points out the important relations that a decision-maker should consider when making an informed decision.

In Figure 7.1 deciding whether or not to launch an attack using an intact nuclear weapon depends on knowledge about the security of the nuclear weapon storage site. The decision also depends on whether the terrorist group can break coded locks on the weapon. Many nuclear weapons are protected by codes such as permissive action links (PALs). If the correct PAL code is not entered into the weapon, the weapon cannot fire. The decision maker would have to consider what capabilities his group has to try to break the code using its own resources or what outside assistance his group could receive, perhaps through coercion or sympathetic aid from officials with knowledge about any codes. Another crucial element for the decision maker is the group's ability to deliver the weapon to the intended target. If any of the factors are too uncertain or impose too great of an impediment, the decision maker could decide to not launch an attack using an intact nuclear weapon.

Influence diagrams can include even more details than the relatively simple example just examined. For instance, the diagram in Figure 7.1 could include a chance variable of the ability to recruit insiders to help obtain the weapon. There

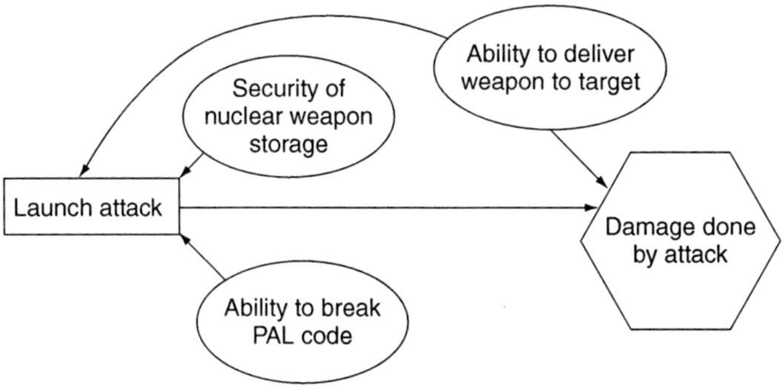

Figure 7.1 An influence diagram of an attack using an intact nuclear weapon.

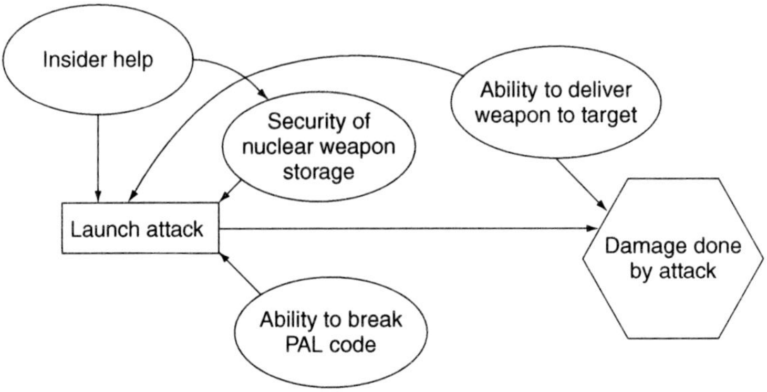

Figure 7.2 An influence diagram of decisions on intact nuclear weapons involving insider help.

would be uncertainty about the effectiveness of the insider and the probability of the operation failing if the insider is arrested by authorities (Figure 7.2).

For a scenario involving a terrorist group making an improvised nuclear device (IND), some different chance variables come into play than for an attack using an intact nuclear weapon (Figures 7.3 and 7.4). In particular, the decision-maker would have to consider the capability of his group to build the device. If his group does not have this capability, he would have to consider whether the group can recruit the necessary technically competent people. The ability to build the IND includes a suite of skill sets such as the ability to work with HEU or plutonium to mold it into the correct shapes for an IND, the capability to obtain and work with high-energy conventional explosives (for example, HMX or RDX) and the ability to acquire or work with the other non-nuclear components of the weapon, including tampers, reflectors and high-speed electronics. Probably the most difficult part of the decision involves assessing where the nuclear explosive material can be acquired. Can a black market be used? What is the likelihood of becoming ensnared in a sting operation? If a trusted source is eventually identified, how much will acquisition of the nuclear explosive material cost? Does the terrorist group have enough resources or money to pay the price?

The construction of an RDD involves similar considerations as the construction of an IND. However, the probability of successfully building an RDD is considered to be much greater than the probability of building an IND. However, the ability to deliver an RDD to an intended target could be more difficult to do for many scenarios than the ability to deliver an IND because the radioactive signature for a potent RDD can be easier to detect than the radiation emitted from an IND. HEU and plutonium in quantities useful for nuclear explosives emit weaker radioactive signatures than the amounts of cesium-137, cobalt-60 or strontium-90 that can fuel potent RDDs. However, radiation detectors currently provide very little or no coverage of cities and other potential targets of radiological terrorism.

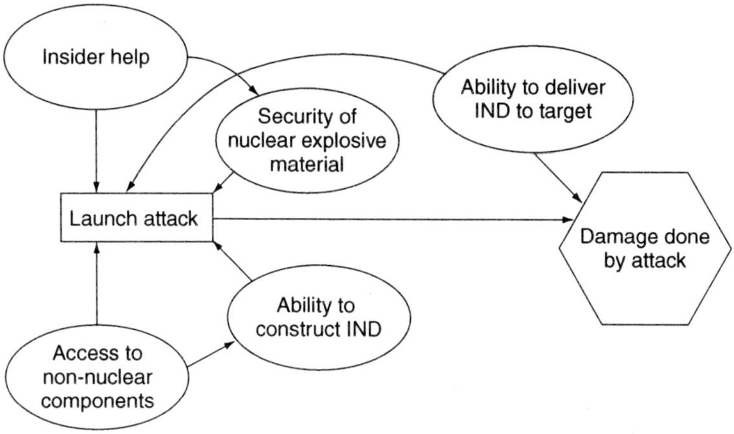

Figure 7.3 An influence diagram of an attack using an improvised nuclear device.

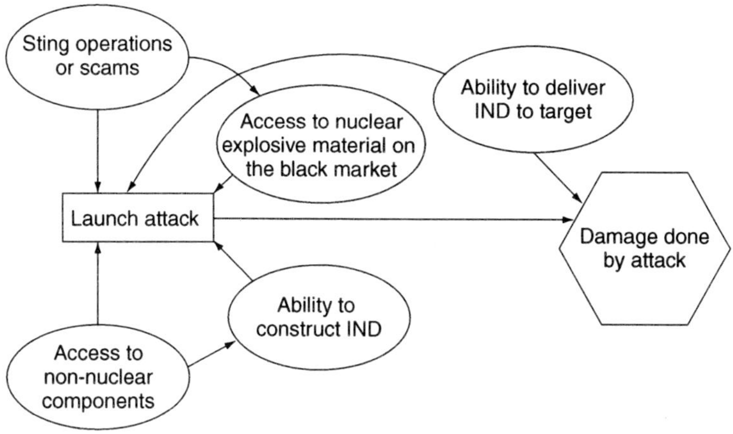

Figure 7.4 An influence diagram of an IND scenario involving the black market and sting operations or scams.

While New York City is in the process of installing rings of radiation detectors as far away as 80 km from the city center, essentially no other American or European city has such radiation detection capability, if any. Major ports have begun to have radiation detectors installed, but terrorists could bypass these ports when transporting powerful radioactive materials. Moreover, even where there is radiation detection coverage, the detectors are far more likely to detect false positives such as radioactive materials in legitimate commercial products (including ceramic tiles and cat litter) than detect radioactive materials meant to be used in acts of terrorism. In sum, radiation detection does not offer a strong line of defense, and the chance of detection, especially outside of some major ports and New York City, appears marginal.

The diagram in Figure 7.5 could also have included access to insider assistance. Such assistance could be a double-edged sword. On the one hand, insiders could give a boost to the operation in its capability to acquire the radioactive material through breaching security at a facility or through informed access to a black market. On the other, insiders could expose the operation to authorities either wittingly or unwittingly. A worrisome variation on the RDD scenario is the concern that a terrorist group could exploit weak regulatory controls by forging licensing documents (Figure 7.6). During the past two years, for example, the US Government Accountability Office, a US government agency that acts as a

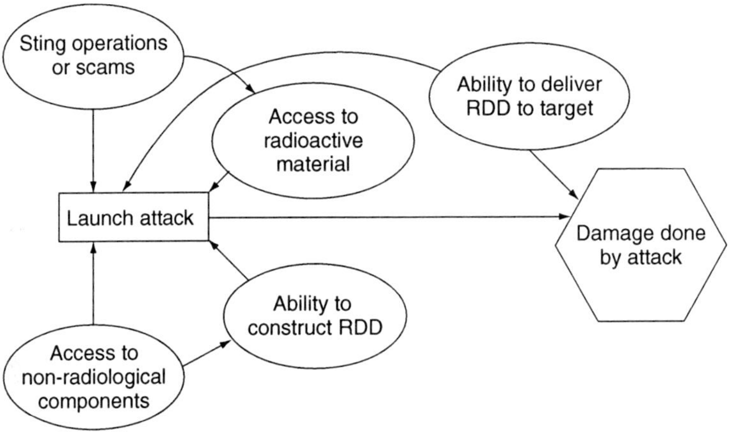

Figure 7.5 An influence diagram of an RDD attack.

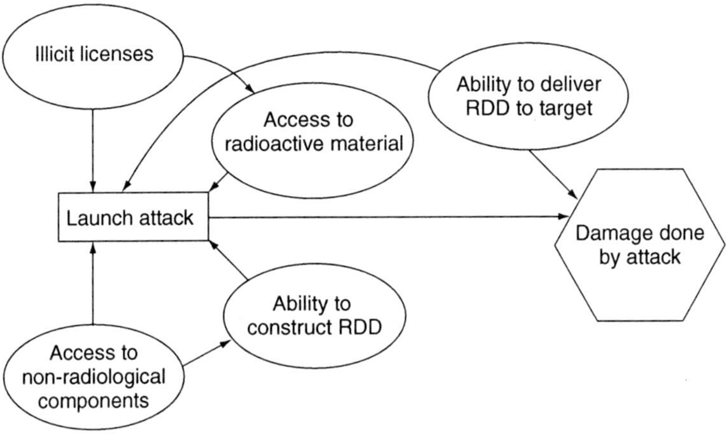

Figure 7.6 An influence diagram of an RDD scenario involving illicit licenses.

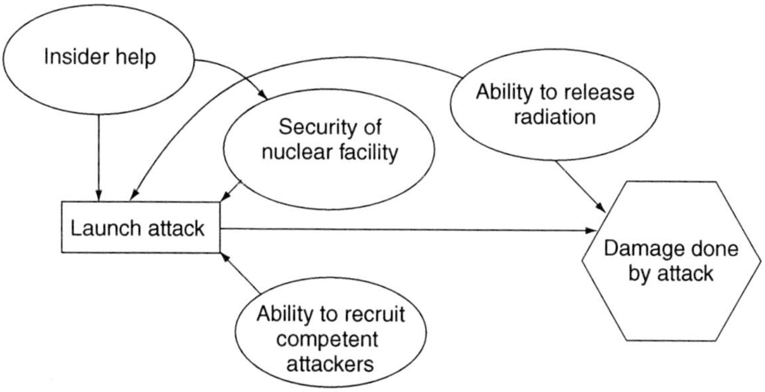

Figure 7.7 An influence diagram of a decision to attack a nuclear facility.

watchdog on waste, fraud and abuse, proved that it could obtain radioactive materials by using illegal licenses.[15]

Turning to attacks or sabotage of a nuclear facility to release radiation, the decision to launch the operation would depend on an assessment of the physical security of the facility, the ability of the terrorist group to breach the security, the robustness of the safety features of the facility to prevent a significant release of radiation and the access to insiders. Even if it were determined that the likelihood of a significant release of radiation were low, the decision to launch the operation might still proceed as long as the decision-maker believed that there was a relatively high likelihood of causing significant economic damage to the facility as well as the economy of the country under attack. Nuclear facilities are a potent symbol of a country's strength. In addition, a terrorist group could seek to cause global damage by causing significant harm to a nuclear plant. With the global buzz about a potential nuclear renaissance, there are concerns that a major accident at, or attack on, a nuclear power plant could have global repercussions.

Indicators of terrorist convergence to nuclear or radiological terrorism

Nuclear and radiological terrorism represent choices among many attack options. How do terrorists choose which attack option to use? Conventional terrorist methods have a long history. Thus, terrorist decision-makers can draw on this experience to make relatively sound predictions about the success or failure of such methods in an attack. In contrast, there is comparatively little history of attempts to commit nuclear or radiological terrorism. Nonetheless, thinking through the decisions that a terrorist leader would have to make points to indicators for the potential terrorist convergence on nuclear or radiological weapons. Many of the indicators of possible terrorist convergence on nuclear and radiological terrorism are indirect and can be considered threshold capabilities that the terrorist group would have to leap over to be able to have a chance at doing these types of terrorism.

A principal indirect critical indicator is whether the group has acquired the human talent necessary to work with nuclear and radiological materials. Has the group recruited nuclear scientists or physical scientists with knowledge of how to use these materials? Similarly, has the group recruited people with technical expertise in incorporating these materials into weapons? Of course, the technical talent required for nuclear weapons is greater than that required for radiological weapons. But the talent required for radiological weapons, especially for those weapons that can effectively disperse radioactive material, is greater than that required to make improvised explosive devices. If that sophisticated technical talent is not available, the terrorist decision-maker could still opt for going ahead with a dirty bomb attack, but his group's chances of causing massive disruption are greatly diminished. Detecting the recruitment of technically talented people can be very challenging. Nonetheless, the skill sets needed for nuclear or radiological weapons can be specified and can thus help guide intelligence community officials to develop ways to monitor where this technical talent can be recruited. For example, universities with student groups that have expressed sympathy or support for terrorist groups might provide opportunities for recruitment of technical talent.

Another threshold indicator is whether the terrorist group has adequate financial resources. Does the terrorist group have the financial assets to buy nuclear weapons, nuclear explosive material or radioactive material? Nuclear terrorism could require upwards of several to tens of million dollars.[16] In contrast, radioactive material would cost far less, thousands of dollars for relatively potent radioactive sources instead. Thus, nuclear terrorism has a much higher financial hurdle to cross than radiological terrorism. Thousands of dollars is a relatively small sum of money, and most terrorist groups would likely have access to this amount. Thus, this indicator is more important in ruling out those groups that cannot cross the threshold of millions of dollars for nuclear terrorism. Authorities would need to monitor terrorists' financial assets to make this estimate for each terrorist group.

A direct indicator of potential terrorist convergence on nuclear or radiological terrorism is the pronouncements of the group's leaders. Are they publishing print or internet writings or making video or audio recordings that express interest in these types of terrorism? A related indicator is whether these groups are researching nuclear or radiological terrorism. Are they exchanging nuclear or radiological weapon designs with like-minded terrorists or extremists? Authorities could monitor websites frequented by these groups. It might be possible to monitor who has checked out certain books such as the *Los Alamos Primer* from libraries or purchased such books from bookstores or the internet. But this type of monitoring of individual book purchases or library use encroaches on civil liberties in many countries. However, it might be possible to monitor whether certain web domains have accessed sensitive nuclear reading materials on the web without revealing individuals' access.

Another direct indicator is whether there is evidence of terrorists probing or surveying nuclear weapons storage sites, facilities containing nuclear explosive material, nuclear power plants or places containing powerful radioactive sources. Such probing could take the form of suspicious activity near these locations. For example, the US National Intelligence Council reported that Russian authorities had detected Chechen rebels performing reconnaissance of nuclear weapons storage sites as well as nuclear weapons transports.[17] Also, it is an indicator of interest if terrorist groups are trying to recruit insiders at nuclear facilities. Detecting such recruitment would likely require personnel monitoring at these facilities, such as personnel reliability programs.[18]

Future work

As even a qualitative decision analysis makes clear, the probability of successful terrorist attack tends to decrease as the complexity of the operation increases. The next stage of the analysis should involve a quantitative assessment of the terrorist group leader's decision process. Although decision analysis has been applied to business applications for about 30 years, it is still relatively new from the perspective of terrorism studies. Recently, there have been a few studies that have applied quantitative decision analysis to counter-terrorism.[19]

A quantitative development of the present study would quantify the probabilities or define probability distributions of the chance variables such as the security of nuclear facilities and the ability to deliver a nuclear or radiological weapon to a target. These probabilities could be estimated by surveying the open literature or polling experts. A quantitative assessment would also factor in the risk acceptance of the decision-maker by defining and introducing into the analysis a mathematical function that models risk-averse or risk-tolerant behavior. Once the probabilities for the chance variables and risk tolerance are defined, the influence diagrams can be transformed into decision trees, which can be used to calculate the expected value of terrorist operations available to the decision-maker. A rational decision-maker would choose the operation with the highest expected value.

Finally, to understand the capabilities of counter-terrorism, the decision analysis can include a second decision-maker representing counter-terrorism or security officials.[20] Additional decision-makers can be introduced representing political leadership and other essential roles. Such multiple agent games can analyze the effect of increasing security on certain facilities and not on others or the effect of sting operations. For instance, what types of sting operations and how frequently should they be run to poison the well for nuclear or radiological terrorism? Such games can lend themselves to assessments of effective deployment of scarce governmental resources.

Notes

1 David C. Skinner, *Introduction to Decision Analysis*, 2nd edition (Gainesville, FL: Probabilistic Publishing, 1999), p. 13.

2 For more technical details on gun-type and implosion-type bombs, see Richard L. Garwin and Georges Charpak, *Megawatts and Megatons: A Turning Point in the Nuclear Age?* (New York: Knopf, 2001) and Charles D. Ferguson and William C. Potter with Amy Sands, Leonard S. Spector and Fred L Wehling, *The Four Faces of Nuclear Terrorism* (New York: Routledge, 2005).
3 Nuclear explosive materials also include americium, neptunium and uranium-233, but these materials are not currently in great enough abundance to pose a significant threat to be used in a nuclear explosive. Conceivably, if India moves toward a thorium-based nuclear fuel cycle, it could make large amounts of uranium-233, a material that could fuel a gun-type bomb.
4 David Albright and Kimberly Kramer, "Fissile Material: Stockpiles Still Growing," *Bulletin of the Atomic Scientists*, November/December (2004): pp. 14–16.
5 US General Accounting Office, *DOE Needs to Take Action to Further Reduce the Use of Weapons-Usable Uranium in Civilian Research Reactors*, GAO-04-807, (2004): p. 28.
6 US National Academy of Sciences, *Management and Disposition of Excess Weapons Plutonium* (Washington, DC: National Academy Press, 1994), pp. 32–3; and U.S. Department of Energy, Office of Arms Control and Nonproliferation, *Final Nonproliferation and Arms Control Assessment of Weapons-Usable Fissile Material Storage and Excess Plutonium Disposition Alternatives* (Washington, C: Department of Energy, NN-0007, 1997).
7 Henry Sokolski (ed.), "Falling Behind: International Scrutiny of the Peaceful Atom," Report of the Nonproliferation Policy Education Center on the International Atomic Energy Agency's Nuclear Safeguards System, July 2007.
8 Michael A. Levi and Henry C. Kelly, "Weapons of Mass Disruption," *Scientific American*, November (2002).
9 Comments by Fire Chief Joseph Pfeifer in Rich Shapiro, "FDNY Chief: We're Training for Nukes," *New York Daily News*, October 22, 2005.
10 Charles D. Ferguson, Tahseen Kahseen and Judith Perera, *Commercial Radioactive Sources: Surveying the Security Risks*, Occasional Paper No. 11 (Center for Nonproliferation Studies, Monterey Institute of International Studies, 2003)
11 International Atomic Energy Agency, "Categorization of Radioactive Sources," IAEA-TECDOC-1344, July 2003.
12 Skinner, op. cit., p. 142.
13 See The National Commission on Terrorist Attacks Upon the United States, *The 9/11 Commission Report*, authorized edition (New York: W.W. Norton & Company, 2004), chapter 5.
14 Steve Coll, "The Unthinkable: Our Shield Against Nuclear Terrorism," *The New Yorker*, March 12, 2007.
15 US Government Accountability Office, "Nuclear Security: Actions Taken by NRC to Strengthen its Licensing Process for Sealed Radioactive Sources are not Effective," Testimony before the Permanent Subcommittee on Investigations, Committee on Homeland Security and Governmental Affairs, US Senate, July 12, 2007.
16 Peter D. Zimmerman and Jeffrey G. Lewis, "The Bomb in the Backyard," *Foreign Policy*, November/December (2006). This article also describes the technical skills needed for the terrorist group to build a gun-type nuclear bomb.
17 US National Intelligence Council, "Annual Report to Congress on the Safety and Security of Russian Nuclear Facilities and Military Forces," December, 2004.
18 Ryan Crow, "Personnel Reliability Programs," Report for Project Performance Corporation, 2004. Available at: www.ppc.com/modules/knowledgecenter/prp.pdf.
19 Steve Eisenhawer, Terry Bott and D.V. Rao, "Assessing the Risk of Nuclear Terrorism Using Logic Evolved Decision Analysis," Los Alamos National Laboratory, LA-UR-03–3467, June 2003; Detlof von Winterfeldt and Terrence M. O'Sullivan, "Should We Protect Commercial Airplanes Against Surface-to-Air Missile Attacks

by Terrorists?" *Decision Analysis*, June (2006): pp. 63–75; Erim Kardes and Randolph Hall, "Survey of Literature on Strategic Decision Making in the Presence of Adversaries," CREATE Interim Report, March 15, 2005.

20 Daphne Koller and Brian Milch, "Multi-Agent Influence Diagrams for Representing and Solving Games," *Proceedings of the 17th International Joint Conference of Artificial Intelligence* (2001).

Part IV

CRBN and terrorism: dilemmas of prediction?

8 Approaching threat convergence from an intelligence perspective

Gregory F. Treverton

The change in targets for intelligence is dramatic. To be sure, for all the emphasis on terrorists and other non-state (or transnational) targets, the change is not absolute. Intelligence dealt with non-states before, and nation-states, like North Korea. Iran or China still will loom large in the work of US intelligence. While some of those state targets, like North Korea, are familiar in the sense that they resemble the secretive Soviet Union, others present different and unfamiliar – if not entirely new – challenges.

This chapter lays out the change in targets, focusing on transnational targets such as terrorists. The change is widely acknowledged, yet its implications run far deeper than is usually recognized. The change goes to the heart of how intelligence does its business – from collection, to analysis to dissemination, to use the labels that are less and less apt. This is nowhere truer than in dealing with the question of terrorists acquiring chemical, biological, radiological or nuclear (CBRN) weapons. For comparison, the chapter first lays out essential characteristics of states as an intelligence target. Then it spells out in some detail the characteristics of transnational targets, then uses CBRN as an illustration of the challenge.

Characteristics of state targets

Questions about states came and come with considerable scope. States are geographic; they come with an address. As important, they come with considerable "story" attached. I have come to think that intelligence ultimately is storytelling. It is helping those who will take action build and adjust the stories in their heads that will guide their decisions. Absent some story, new information about a topic is just a factoid. The story provides a pigeonhole and context for that new information. To be sure, if the story in people's heads becomes too hardened, too impervious to adjustment in light of new information, that new information may simply be discarded if it doesn't conform to the story. This is called mindset or "groupthink" and is the root of most of what are referred to as intelligence failures.

We know what states are like, even states as different from the United States as the Soviet Union or North Korea. They are hierarchical and bureaucratic.

Because their purposes are broadly similar – protecting and seeking to assure decent living conditions for (at least some of) their citizens, their internal institutions share similarities. The devil may be in the details in understanding, for instance, the differences in the workings and influence of the Japanese Diet by comparison to the US Congress, but at least "legislature" provides a shared starting point for the discussion. So, the fact that states come with a story attached not only facilitates the task of analyzing them, it also greases the conversation between intelligence and policy about the results of that analysis.

Most intelligence questions about states fell, and fall, into the frequently used distinction between *puzzles* and *mysteries*.[1] Puzzles are questions that could be answered with certainty if only with access to information that is, in principle, available. Dave Snowden, in his business-related writing about the philosophy of information, calls them *known problems*, for which there is a unique relationship between causes and effects.[2] The challenge is to correctly categorize the problem, obtain the necessary data to solve it, and apply accepted formulas. Military targeting issues are puzzles. So are many issues about state's capabilities. How many nuclear devices does North Korea have?

Much of Cold War intelligence was puzzle-solving, looking for additional pieces to fill out a mosaic of understanding whose broad shape was a given. Because so much that was open about the United States and other democracies was secret about the Soviet Union – basic economic and military statistics, for instance – the United States and its allies spent billions of dollars on exotic collection systems to solve those puzzles.

Because those puzzles were secret, by definition, solving them relied heavily on secret intelligence sources – espionage (or human intelligence, HUMINT) and what came to be called "technical collection" by "national technical means," primarily satellites taking pictures and other imagery (imagery intelligence, or IMINT) or satellites, ground stations and other platforms intercepting signals from those states (signal intelligence, or SIGINT).

HUMINT, especially, is nicely matched to puzzle-solving. HUMINT is, by nature, a target of opportunity business. A spy in place may not be invited to a critical decision meeting, or that spy may attend but not be able to get the information to his or her handler quickly lest the spy's cover be blown. As a result, spies may or may not be able to provide useful information about fast-moving plans or intentions. For most puzzles, however, the puzzle piece that isn't available today will still be welcome tomorrow. So if the spy can't steal that secret puzzle piece today, it will still be valuable if provided tomorrow. The result is an apparent paradox: while spying is often conceived as a way to get an inside view of an adversary's intentions, and sometimes is, it is more reliably valuable at solving puzzles. It is little surprise, then, that most espionage targets are puzzles – military plans, weapons designs or industrial processes.

In contrast to puzzles, no evidence can solve mysteries definitively, for typically they are about people, not things. They are contingent. Snowden labels these *knowable problems*, which involve contingent relationships between a limited set of causes and effects. In this realm, analytic techniques can be used to predict outcomes, at least probabilistically. We cannot know the answer, but we can know – in the case of intelligence, usually from recent history – which factors are important to monitor and something of how they interact to produce the answer. Russia's inflation rate for this year is a mystery. So is whether Israel might strike Iranian nuclear facilities. Notice that while Soviet capabilities were primarily a puzzle, those of terrorist groups are a mystery because they depend until the very last moment on the actions and vulnerabilities of their foes. More on this follows.

For puzzles, the product is *the* answer: North Korea has X nuclear weapons; Soviet missiles have Y warheads. To be sure, the answer may not be definitive; it may remain a best estimate. If that is so, however, it is so not because of inherent uncertainty. Rather, it is so because information that is available in principle is unavailable in fact. The answer is as close at we can come with the information at hand. Notice that the infamous October 2002 National Intelligence Estimate (NIE) about whether or not Iraq had weapons of mass destruction (WMD) sought to solve a puzzle: Iraq either had or didn't have WMD programs, with relevant proscribed materials to match. The NIE has been rightly criticized on a number of grounds. Yet while its answer, a resounding "yes, Saddam has them," was dead wrong in fact, it was probably the only answer that could have been offered given the evidence and argument at hand.

For mysteries, the product is a best forecast, perhaps in the form of a probability, with key factors identified, as well as how they bear on the estimate. If analysts do not and cannot know what Russia's inflation rate will be this year, they do know from Russia's experience as well as that of other countries, both what factors will be important in determining that rate and, at least roughly, how they will combine to produce it. So the answer can be conveyed by laying out those determinants, along with quantitative or qualitative assessments of where they stand and how they are moving. This can lead to a best forecast, along with some bounds of uncertainty. To sharpen both the forecast and the bounds, sensitivity analyses (for instance, using scenarios or a variety of quantitative methods) could test the effects of different levels of determinants on outcomes.

Looking at transnational issues: how are they different?

As an intelligence challenge, transnational targets, like terrorists, differ from traditional state targets in nine main ways. Table 8.1 summarizes those differences:

Transnational targets are not new for intelligence, for it has long been active against organized crime and drug traffickers. What *is* new is twofold – first, the importance of transnational threats, especially – but not just – terrorism; and second, the range of transnational threats of concern. In the past, organized crime and drug traffickers were a secondary activity for intelligence; now, some of the transnational threats are primary.

Moreover, the range of current and prospective transnational targets is very broad. If "threat" is conceived broadly, then threats can be thought of as covering

Table 8.1 From Cold War targets to Era of Terror targets

	Old: Cold War	*New: Era of Terror*
Target	States, primarily the Soviet Union	Transnational actors, also some states
"Story" about target	Story: states are geographic, hierarchical, bureaucratic	Not much story: non-states come in many sizes, shapes
Location of target	Mostly "over there," abroad	Abroad and at home
Consumers	Limited in number: primarily federal, political, military officials	Enormous numbers in principle: including state, local and private
"Boundedness"	Relatively bounded: Soviet Union ponderous	Much less bounded: terrorists patient but new groups and attack modes
Information	Too little: dominated by secret sources	Too much: broader range of sources, though secrets still matter
Interaction with target	Relatively little: Soviet Union would do what it would do	Intense: terrorists as the ultimate asymmetric threat
Form of intelligence product	Intense: terrorists as the estimate with excursions for mysteries	Perhaps "sensemaking" for complexities
Primacy of intelligence	Important, not primary: deterrence not intelligence rich	Primary: prevention depends on intelligence

a range. At one end are those threats that come with threateners attached, people who mean us harm.[3] At the other end are developments that can be thought of as *threats without threateners*. If they are a threat, the threat results from the cumulative effect of actions taken for other reasons, not from an intent that is purposive and hostile. They might also be called *systemic* threats. Those who burn the Amazon rain forests or who try to migrate here or who spread pandemics here do not necessarily wish Americans harm; they simply want to survive or get rich. *Their* self-interest becomes a threat to *us*.

Interestingly, the main transnational targets for intelligence in the past, organized crime and drug traffickers, fell somewhere in the middle of that continuum. The activities of both created national security concerns, but the danger the traffickers posed to the nation was a by-product of their main purpose, which was to enrich themselves. Also in the middle of that continuum, energy imports are commerce, but the products are so crucial to national economies as to invoke security concerns – sometimes stunningly mis-posed. Those concerns have waxed and waned with price and supply, and they have been reconfigured recently not just by much higher prices, but also by the new winners and new losers that result, and by the changes in the nature of global energy trade. Moreover, as both trade and investment become global, economic transactions driven by commerce – with no threat intended – may come to look menacing. So it was in recent years with China's proposed purchase of UNOCAL, or a Dubai-based company's interest in taking over US ports.

What is striking now is that the ends of the continuum are more important. They are primary, not secondary. At the "they mean us harm" end, terrorism by Islamic extremists does not pose the existential threat to the United States that the Soviet Union and its nuclear weapons did, but it is more than an inconvenience. And it frightens Americans well beyond its actual harm so far. For instance, polls asking how concerned respondents were about more terrorist attacks in the United States recorded 49 percent as worrying a "great deal" and 38 "somewhat" immediately after the 9/11 attacks. Two years later those concerned a great deal had fallen to 25 percent, but those "somewhat" concerned was 46 percent.[4]

In fact, the loss of life on 9/11 was off the charts of American's historical experience with terrorism. In the years after 2001, the yearly numbers of fatalities of Americans from terrorism, worldwide, were all less than 100, and usually barely into double digits. Fifty-six was the total for 2005, and that may have resulted as much from a change in counting rules as from an actual increase in fatalities.[5] This compares with an average over the last five years of 62 people per year killed by lightning, 63 by tornadoes, 692 in bike accidents and a whopping 41,616 in motor vehicle-related accidents.

One reason for the fright is that people find it very difficult to deal with low probability occurrences, especially ones with devastating consequences. A generation ago, researchers compared the assessments of experts on nuclear power with those of non-specialist citizens. Two groups of citizens ranked the risk first in a list of 30 activities and technologies, and the third group ranked it eighth. By contrast, the experts ranked it twentieth.[6] For nuclear weapons and nuclear war, the gap was even wider. The citizens may have been right and the experts wrong – happily, we had no proof during the Cold War for nuclear weapons. Working every day on the issue may have desensitized the experts, or they may have reduced their own cognitive dissonance by subconsciously downgrading their own assessments of the risk. In any event, the difference was striking, and on issues like nuclear power, the experts' views did tend to correlate with actual evidence, like numbers of fatalities.

When I spoke about the terrorist threat, especially in the first years after 2001, I was often asked what people could do to protect their families and homes. I usually responded by giving the analyst's answer, what I labeled "the RAND answer." Anyone's probability of being killed by a terrorist today was essentially zero, and would be the same tomorrow, barring a major discontinuity. So they should do: nothing. Not surprisingly, the answer was hardly satisfying, and I did not regard it so. I came to realize what studies of other risks indicate, that terrorism is especially frightening because it seems random.[7] We have the sense that we can do something about other things that might kill; we can diet, or not smoke, or drink less. The sense is partly illusion, but terrorism seems entirely random. It is not – living outside major cities would reduce a person's already vanishingly low probability of being killed. But so it seems.[8]

At the other end of the continuum, global warming seems more and more serious as a global threat, if not directly to the United States then to countries and regions in whose stability it has important interests. And global pandemics are a direct threat to the United States. Like terrorism, global warming runs beyond the mystery category because we are sure neither of the factors that produce it nor, especially, how they combine. Moreover, effects may cascade in discontinuous ways. Like bridges that are not maintained, there may be little visible effect for years, but a catastrophic one all of a sudden. The best intelligence can do may be to provide "if, then" assessments – if warming has given physical effects in particular places, then it will exacerbate given sources of social and political tension.[9]

No "story"

While state targets of intelligence will remain – Iran, North Korea, China or Russia, for example – the shift to terrorists and other transnational developments as primary targets is momentous. States, even states very different to the United States, came with some "story" attached. Transnational targets, especially terrorist groups, do not. These new targets deprive intelligence and policy of a shared story that would facilitate analysis and communication. If we know that states are geographical, hierarchical and bureaucratic, there is no comparable story for non-states, which come in many sizes and shapes.

Take al-Qaeda as an example. More than a half decade after September 11, it was unclear whether the organization was a hierarchy, a network, a terrorism venture capitalist or an ideological inspiration. No doubt it contained elements of all four, and other characteristics, in different measures. But that hardly amounted to a story. As a result, interpreting additional information and communicating those assessments were both difficult. The "hierarchy" and "network" views made it hard to conceptualize the March 2004 Madrid bombers, who seemed to have very little to do with al-Qaeda "central." In turn, once the "inspiration" view of al-Qaeda became accepted, that made for very different interpretations of the extent to which al-Qaeda central had been able to reconstitute itself five years and more after September 11.

Disease has some story attached. We know what diseases are and something of how they spread. Yet, a pestilence brought into the United States might be either a purposive terrorist tactic or an "innocent" contagion due to greater global connectedness and the multiplicity of previously unknown pathogens. As with problems of cybersecurity, the difference is by no means likely to be clear. Today, an opponent can wage a "war without fingerprints," and as a result not only might it be unknown who is attacking us, but that we are being attacked in the first place. Talk about the absence of story!

Location of the target

Transnational targets not only come without a story, by definition they come without an address. They are, after all, transnational. There are both "over there" abroad, and "here," at home. By contrast, while at times early in the Cold War the United States worried about communists at home, for the most part the intelligence challenge of communism could be divided into foreign intelligence abroad and counterintelligence at home. The task at home was mostly following good white-collar spy-masters from the communist countries as they tried to recruit spies in the United States. Dividing the task in that way enabled the distinctions or "oppositions" that shaped the US approach – intelligence versus law enforcement, foreign versus domestic, public versus private. Those distinctions were reflected in organizations and their operating concepts. They made a good deal of sense during the Cold War, especially in safeguarding the civil liberties of Americans. However, transnational threats, especially terrorists, respect none of them. The oppositions set the country up to fail on September 11.

The fact that transnational targets are here as well as over there impels nations to reconsider what has been regarded as domestic intelligence. Not only do they face the necessity of collecting more information on their citizens and residents, and trying to do so with minimal damage to civil liberties, they also need to rethink the boundaries that the distinctions established. Terrorists are criminals but may commit but only a single crime, and then it will be too late. Terrorist groups have a lot in common with international crime organizations except one critical particular: criminals want to live to steal another day; they are not candidates for suicide bombers.

More and more diverse consumers

Not only did consumers share with intelligence a story about states – territorial and usually hierarchical, with histories, traditions, bureaucracies and standard operating procedures – but those consumers were relatively few in number during the Cold War. Most of them were located at the apex of the national security decision-making establishments. They were mostly political–military officials. Busy and distracted, they were not always an easy target for intelligence to serve. They were, however, an identifiable and familiar target. When the Cold War ended, intelligence looked for new consumers and at first found them in domestic agencies like Commerce, which wanted staff work as much as intelligence analysis.

Then, after the 9/11 attacks, intelligence's consumers mushroomed, this time including state and local officials and private managers of infrastructure. For transnational issues, the variety of consumers is also much greater, extending to a variety of agencies outside the traditional national security establishment. There are 18,000 relevant government jurisdictions in the United States, comprised of more than 600,000 police; and nation-wide, there are three times as many private security officers as public police. Virtually none of them have security clearances to see any classified federal information, let alone the secret compartmented intelligence (SCI) that is intelligence's pre-eminent stock-in-trade. Both the FBI and Department of Homeland Security have made efforts since September 11 to clear state and local officials, but the numbers, while significant, remain small. The FBI has provided clearances to roughly 2,000 state and local officials per year, mostly at the secret level.[10]

So, too, the lack of a story for transnational targets like terrorists, or the presence only of tentative, competing stories about them, often based only on short histories, is compounded by the different organizational cultures involved. Traditional intelligence and law enforcement are very different enterprises. Even the meaning of "intelligence" differs between them. State and local officials are unfamiliar with the products of national intelligence agencies, and are prone to imagine (or hope) there is magic behind the green door of classification if only the Feds will open it.

Moreover, who is the "key" consumer is far less clear. An airport security officer or a public health doctor may have a more urgent "need to know" about a threat than the President of the United States because he or she may be in a more immediate position to thwart it. At the same time, most new transnational consumers have little familiarity with intelligence and how to interpret it. For this reason, too, the challenge of connecting with the consumer is greater in the transnational than in the state-to-state arena.

Less "bounded"

States are, for the most part, clearly "delineated," having known borders and capitals, and doing much that is openly observable. Moreover, states act within the context of formal and customary rules, such as established military doctrines, giving predictability to many of their actions. By contrast, transnational actors are amorphous, fluid and hidden, presenting intelligence with major challenges simply in describing their structures and boundaries. Because such actors are also far less constrained by formal rules than their state counterparts, they can engage in a wider variety of tactics on a regular basis, adding immensely to the challenge of forecasting their behavior.

That relative lack of constraints is obvious in the case of the Islamic extremist terrorists. It is also the case for organized crime and drug trafficking, actions which by definition put their perpetrators on the other side of the law. The political scientist James Rosenau refers to the transnational actors as "sovereignty-free."[11] For him, they confront an "autonomy dilemma" parallel to the "security dilemma" that states face. Yet, free of sovereignty, they also are free of many of the obligations and constraints of sovereign states. Even the transnational actors most of us are likely to label "good" may be in a position to buy themselves considerable freedom of action, for better or worse. Indeed, the scale of some private activities literally outstrips government: the approximately $800 million that the Gates Foundation contributes every year for global health approaches the annual budget of the World Health Organization and is comparable to the funds given to fight infectious disease by the US Agency for International Development.

While the current Islamic extremist terrorists hardly act quickly, but instead plan their attacks carefully over years, transnational targets are in another sense also less bounded than state-centric ones. There will be discontinuities in targets and attack modes, and new groups will emerge unpredictably. Intelligence had a decade to explore the impact of Gorbachev's accession on the Soviet system, but in the case of al-Qaeda, for example, events unfolded at a stunning pace after September 11. Moreover, the transnational arena involves networked actors subject to what students of the emerging science of networks refer to as "cascades," making them more vulnerable to sudden change than state-to-state systems. Some networks, like power grids, are tightly coupled systems. Small changes within the network accumulate until the network reaches a "tipping point," after which dramatic domino-like sequences ensue, such as the collapse of the network – the failure of the electric power grid. To take an example outside of terrorism, in the Asian financial meltdown of 1997 a relatively small crisis involving the Thai currency quickly enveloped much of Asia, as networked financial markets produced cascading effects.[12]

Broader and lower quality information base

Given closed foes, Cold War intelligence gave pride of place to secrets – information gathered by human and technical means that intelligence "owned." Terrorists are hardly open, but an avalanche of open data is relevant to them: witness the September 11 hijackers whose true addresses were available in California motor vehicle records. During the Cold War, the problem was too little (good) information. Now, it is too much (unreliable) information. Then, intelligence's secrets were deemed reliable; now, the torrents on the web are a stew of fact, fancy and disinformation.

Because of the unbounded and high-profile nature of transnational threats, intelligence must wade through a sea of information that contrasts sharply with the much more limited information that was available on closed societies such as the Soviet Union. Much of the information is, at best, of uncertain reliability. Moreover, as compared with a state with a long history, much less contextual information is available that can be used to evaluate the reliability of new information. For these reasons, the problem of separating "signals" from "noise" is more acute in the transnational domain.

That lack of context – the lack, in other words, of a story – means that information-gathering against terrorists necessarily involves "mining" or other processing of large quantities of information. After the September 11 attacks, with the names of the hijackers known, the government could quickly pick up their trail through motor vehicle records, addresses, credit cards and the like. But that was after the fact. Before the fact, names of interest may yet be unknown, or following them bedeviled by aliases or by different transliterations of the same Arabic name.

Whatever the legal debate over the National Security Agency's post-September 11 Terrorist Surveillance Program, the program vividly illustrated the challenge of dealing with the presence of vast amounts of information and the absence of much context for processing it.[13] In the wake of September 11, the US government worried about new plots and new cells, but it had few specific leads. Rather, it trolled through large numbers of phone calls, almost all international, where there was reason to believe one of the participants had links to al-Qaeda – for instance, because the call came from a region of Afghanistan or Pakistan where al-Qaeda was thought to operate.

Interaction with target

Former US Secretary of Defense Harold Brown is said to have quipped about the US–Soviet nuclear competition, "When we build, they build. When we stop, they build." While various countries, especially the United States, hoped that their policies would influence Moscow, as a first approximation, intelligence could presume that they would not. The Soviet Union would do what it would do. The challenge, in the first instance, was figuring out its likely course, not calibrating influence that other nations might have over that course.

The terrorist target, however, is utterly different. It is the ultimate asymmetric threat, shaping its capabilities to our vulnerabilities. The September 11 suicide bombers did not hit on their attack plan because they were airline buffs. They had done enough tactical reconnaissance to know fuel-filled jets in flight were a vulnerable asset and defensive passenger clearance procedures were weak. They could get box cutters through airport security, and the scheme obviated the need to face a more effective defense against procuring or importing ordnance. By the same token, the London, Madrid and other bombers did enough tactical reconnaissance to shape their plans to the vulnerabilities of their targets.

To a great extent, we shape the threat to us; it reflects our vulnerable assets and weak defenses. In that sense, the capabilities of terrorists are a mystery, not a puzzle, for those capabilities depend on their adaptation to the vulnerabilities of their targets, not on counts of missiles, guns or even cells. For instance, al-Qaeda-linked plotters in 2006 planned to blow up airplanes over the Atlantic with liquid explosives smuggled onto planes as sport drinks or other permitted carry-ons. They had adapted to the airport security procedures then in effect, knowing that drinks were then permitted as carry-ons and that most detectors in place could not identify explosives.

As military planners would put it, it is impossible to understand red (potential foes) without knowing a lot about blue (ourselves). In contrast to states like the Soviet Union, transnational actors like terrorists have a more intense relationship with the dominant actor in the international system, the United States. Their tactics are often predicated upon our policies and defensive measures, making their behavior less determinate and predictable. Our understanding of transnational actors' proclivities will lead us to take actions that will – to a greater extent than would be the case with more structured, internally driven state actors – prompt adaptive behavior on their part. This process of adaptation can turn the predictions of intelligence into "self-negating prophecies" – and that is if the intelligence is good enough to make accurate predictions.

This interaction between "us" and "them" has very awkward implications for intelligence, especially foreign intelligence that has in many countries been enjoined from examining the home front and, less formally, worried that getting too close to "policy" is to risk becoming politicized. The task for intelligence now cuts directly across the foreign–domestic distinction. Moreover, to the extent that intelligence now becomes a net assessment of red against blue, that too, is something that has been the province of the military, not civilian agencies.

"Sensemaking" of "complexities"

Transnational targets involve both puzzles and mysteries, but they also invoke what might be thought of as *mysteries-plus*, what Snowden calls "complexities." Where al-Qaeda leaders hid out along the Pakistan–Afghanistan border was a puzzle, albeit one whose solution might have altered rapidly as the leaders shuttled among hide-outs. When, where and how al-Qaeda might attack the United States might be thought of as a mystery. It can be so conceived to the extent we believe we have some understanding of the nature of al-Qaeda and its links to affiliates, of its strategy, proclivities and so on – to the extent, in other words, that there is now some "story" about al-Qaeda.

Yet the mystery conception seems not quite apt as a way to frame the challenge of understanding Islamic extremist terrorism and some other transnational targets. Because that terrorism comes with relatively little relevant history and context, and because it is so unbounded, it involves a wide array of causes and effects that can interact in a variety of contingent ways. Large numbers of relatively small actors respond to a shifting set of situational factors. Moreover, because interactions reflect unique circumstances, they do not necessarily repeat in any established pattern and are thus not amenable to predictive analysis in the same way as mysteries. Both the 9/11 terrorists in 2001 and the Fort Dix plotters in 2006–2007 had connections to al-Qaeda, but the links were very different, and the first did not provide any sense of pattern for the second.

To be sure, the distinction between transnational and traditional intelligence problems should not be overstated: there are some state-to-state problems, such as battlefield situations or crisis diplomacy, where situationally driven interactions among a large number of players can also produce a wide variety of outcomes.

These complexities in understanding transnational issues require combinations of regional and functional expertise. While a country political or economic analyst can often work in relative isolation from analysts with other specializations, that is not so for terrorism. Weapons proliferation analysis, for instance, draws upon specialists in science and technology, illicit transfers, money laundering, politics and network behavior, to name but a few, to track and comprehend the activities of weapons networks. To a much greater extent than in traditional areas, transnational analysis is a team or even networked activity (as specialists will be located in many agencies). This has both potential benefits in terms of avoiding mental biases (mixing different perspectives), but also potential risks in the form of groupthink and "lock-in."

The challenge of dealing with complexities still lies ahead for the most part. The goal is to convey a sense of emerging patterns with an eye to reinforcing or disrupting, respectively, positive or adverse patterns. Communicating that sense may be hard to do in a discrete paper or stream of electrons. Rather, it may be best done with the active participation of policy officials – for instance, using computer power to *fly through* a wide range of variables and scenarios, looking for patterns.[14] Such processes run into two familiar obstacles – the canonical separation of intelligence from policy, and the fact that policy officials, in particular, are always hard-pressed for time. Because the product is a sharpened sense, the problem is not one simply of communication; the sense needs to develop out of shared analytic work.

The product is what might be called *organizational "sensemaking,"* as developed particularly by the noted organization theorist, Karl Weick.[15] Sensemaking is the process through which organizations – not individuals – comprehend the complex environment with which they must contend. It is a continuous, iterative, largely informal effort to understand, or "make sense" of what is going on in the external environment that is relevant to the organization's goals and needs. In essence, it is the collective intuition of an organization. Through conversations at all levels, organizations construct ongoing interpretations of reality by comparing new events to past patterns, or in the case of anomalies, by developing stories to account for them. Weick argues that the fluid sensemaking process has clear advantages as a framework for organizational action over "decision-making" because the latter often locks the organization into polishing and defending formal decisions that may no longer be appropriate in fast-changing situations.

The methodology for sensemaking – creating a unified, explanatory, consensual understanding about the world that leads to principled, consistent action – requires forms of analysis and interactions with consumers that are not yet developed, well known, or widely used. Because it requires common understanding, it probably requires more interactive intelligence process than products, and so there is a premium on ways to facilitate that interaction. After all, the real need is not good analysis on a piece of paper – too often in intelligence, paper goods (or streams of electrons) are treated as ends in themselves; rather, the need is for improved understanding in the heads of officials who will decide or act.

It is harder to spell out sensemaking in practice because it is not yet fully developed or implemented; it remains largely a theoretical construct. However, return again to the question of when, where and how al-Qaeda and its kin might next strike the United States. Viewed from a sensemaking perspective, the challenge would entail a continuing conversation between intelligence and policy, one conducted in the knowledge that the sensemaking process could fail (or succeed) dramatically at any moment. The conversation would be as informal as possible, aiming to sustain open-mindedness about a wide variety of terrorist organization and connections, motives and attack modes. It would test new information against a wide variety of hypotheses – in the absence of high confidence about which factors mattered and how they were connected – and to create new ones as necessary. The product would be a sharpened sense for possibilities and probabilities, against the understanding that a high degree of uncertainty was ineradicable. The CBRN illustration later in this chapter will try to sharpen the sense for sensemaking.

Importance of intelligence

The last major difference between transnational targets, especially terrorists, and state targets like the Soviet Union may be the most important of all. If *prevention* is the name of the game, the pressure on intelligence is extraordinary. In his 2002 national security strategy, President Bush was speaking of Iraq but was graphic about the need to prevent attacks, by preemptive action if need be:

> We must be prepared to stop rogue states and their terrorist clients before they can threaten or use weapons of mass destruction against the United States and our allies and friends .... To forestall or prevent such hostile acts by our adversaries, the United States will, if necessary, act pre-emptively.[16]

Or, as he put it more colorfully in his speech to the nation on March 19, 2002:

> We will meet that threat now, with our Army, Air Force, Navy, Coast Guard and Marines, so that we do not have to meet it later with armies of fire fighters and police and doctors on the streets of our cities.[17]

He had foreshadowed the new strategy in his speech at West Point in June 2002: "By confronting evil and lawless regimes, we do not create a problem, we reveal a problem. And we will lead the world in opposing it [*sic*]."[18]

In contrast, the dominant strategy of the Cold War, deterrence, was not so sensitive to the specifics of intelligence. It rested on the assumption that, for all its differences in goals and ideology, the Soviet Union was like us – modern, rational (in our terms) and not self-destructive. It took Soviet intentions as hostile but rational. Thus, once Moscow had nuclear weapons, the way to ensure that it didn't use them was deterrence, which came to be associated with second-strike retaliation: so long as the United States (and its allies) had the capacity to wreak unacceptable damage on the Soviet Union after a Soviet nuclear first strike, that first strike would never come and the promise of US retaliation would never be tested.

Two of the four main lines of debate about Cold War deterrence didn't much involve intelligence, and one that did was highly technical. The first was: how much is enough? That is, how much second-strike destruction would be enough to deter any Soviet first strike? The second was the question that animated US–European discussions during the Cold War: if NATO judged itself dangerously inferior to the Warsaw Pact in conventional forces along the central front in Europe, how could it credibly invoke some possibility of using nuclear weapons first if a Pact conventional attack in Europe were succeeding? That sometimes led to torturous reasoning and tortured policy, but those policy issues didn't much turn on intelligence.[19]

The logic of deterrence did raise one intelligence issue, a technical one invoked by the awful, paradoxical aphorism about deterrence: offense is defense and defense is offense; killing people is good, killing weapons is bad. By this logic, if the Soviet Union had high confidence that it could defend against a retaliatory strike – by some combination of killing US missiles in a first strike, then killing the residual missiles en route to the Soviet Union – it might be tempted by a first strike. The challenge for intelligence came down especially to assessing critical puzzles: how many independent warheads Moscow had, with what accuracy, and thus how much of a threat the Soviet Union posed to US missile forces for retaliation.

The fourth issue might have involved intelligence but for the most part did not. That was Soviet *intentions* and *perceptions:* how much risk of nuclear war was the Soviet leadership willing to run, and how much did it see the nuclear stand-off in terms akin to those of the United States? To be sure, these issues were hotly debated, and intelligence bore on them. But they were mysteries at least, and so no intelligence was very decisive. The only exception was one episode during the Reagan build-up of the 1980s, when Moscow, imprisoned in its own paranoia, had convinced itself that the United States was preparing a nuclear first strike against it. Fortunately, the KGB's chief officer in London had volunteered to spy from Britain. He alerted British officials to Moscow's concern, and Britain shared his intelligence with the United States. Senior American leaders, including President Reagan, were aghast and took pains to reassure their Soviet counterparts.[20]

In contrast to deterrence, prevention – whether by preemption, disruption or simply defending vulnerabilities – requires enormous precision in intelligence. Consider military preemption against enemies' dangerous weapons, like CBRN. America's capacity for "ISR" – intelligence, surveillance and reconnaissance – is unparalleled, in a world class by itself. It is also improving rapidly. However, its shortcomings are virtual descriptions of features of foes' CBRN programs. Existing ISR is not good at detecting objects that are hidden under foliage or concealed or, especially, underground. Nor is it good at locating objects precisely by intercepting their signals. Would-be proliferators will take pains to conceal their facilities or change the pattern of activities at weapons sites, as India did before its 1998 explosion of a nuclear weapon. With respect to the puzzle of North Korean nuclear weapons, the best US intelligence could do in the 2000s was a guesstimate that the county had a few but with no judgment about location anywhere good enough to permit a preemptive strike.

If terrorist threats are to be prevented by disrupting them or by closing the vulnerabilities they seek to exploit, that puts immense pressure on intelligence to understand threats well enough, soon enough. It means moving back up the chain from possible terrorist acts to groups and their proclivities, if not their intentions. It requires an understanding of mysteries or complexities that is as fine-grained as the Cold War's understanding of the puzzle of Soviet missile capabilities. In one respect, though, deterrence and prevention are alike: in both cases, the goal is nothing – no attack. That raises the question of metrics – for deterrence, metrics for policy more than for intelligence, and for prevention, for intelligence more than for policy because prevention is so intelligence intensive. If there is no attack, did policy and intelligence succeed, was the country merely lucky or was the threat exaggerated from the beginning?

"We're all Bayesians now"

I bumped into a former colleague, a Soviet, now Russian, specialist, recently in the halls of the CIA, and we had a quick conversation about methods of intelligence analysis. At one point, he quipped: "We're all Bayesians now." The quip stuck with me. "Bayesian" derives from a famous theorem discovered by an English preacher, Thomas Bayes and published in 1763, and has come to describe both an inclination and some process to update subjective probabilities in light of new evidence. In an important sense, almost all intelligence analysis is Bayesian and always has been, for even with regard to puzzles, finding *the* piece that will solve the puzzle with certainty is rare. However, the uncertainty and un-boundedness of transnational targets, especially terrorists, has underscored the need for both a Bayesian attitude and for some more formal approaches to making that approach concrete.

The basic idea is easiest to capture with a puzzle. Suppose I hand you a coin. You assume that it's fair and that if you toss it, heads and tails are equally likely outcomes. So your initial estimate is that it's a fair coin. But suppose you begin tossing it. If you get heads three times in a row, you might begin to suspect the coin is not fair but probably wouldn't reject that assumption altogether, for you know that the chances are one in six that a fair coin would yield three straight heads. What you would do is adjust your initial certainty that the coin was fair, beginning to entertain the thought that it might not be. If additional tosses produced still more imbalances toward heads, you'd move toward the judgment that the coin was unfair.

In a Bayesian sense, Donald Rumsfeld's famous quote about UN inspections searching for Iraqi WMD in the run-up to the 2003 war was dead wrong. What he said is: "The absence of evidence is not necessarily the evidence of absence."[21] What he meant was that just because the UN inspectors hadn't found evidence of active Iraqi WMD programs didn't prove that Iraq had no such programs. And of course that was right. Yet in a Bayesian sense, each day the UN inspectors did not find evidence of WMD should have shifted the odds a little in the direction of Iraq not having them. The way the puzzle was framed – does Saddam Hussein have WMD? – meant that it could only have been solved definitively in one direction, yes he does. Otherwise, Rumsfeld could have continued to repeat the same argument until every square inch of Iraq had been searched. But in a Bayesian sense, he would have been wrong not to entertain more and more strongly the possibility that Iraq, in fact, had no WMD programs or stockpiles.

The Bayesian approach is easiest to grasp for puzzles because of the direct relationship between new puzzle pieces of information and more confidence about the puzzle's solution. But as an approach, a way of thinking, it is probably more important still for mysteries and complexities, for it not only underscores the inherent uncertainty but also forces hard conversation – in a sensemaking process – about precisely how a new piece of evidence (or logic) cuts and how much difference it makes. The illustration of terrorists seeking a CBRN weapon will make clearer the process and its Bayesian character.

In one sense, the CBRN problem is no different than many more familiar intelligence challenges: it requires intelligence from several sources to be laid atop one another, perhaps in short order – what has come to be called "multi-INT" in intelligence jargon. In this case, as will be seen later, the paths to a CBRN weapon include too few "unique" signifiers – that is, actions that *only* (or virtually only) are required for weapons, not legitimate purposes. Thus, intelligence would have to combine some information about groups with the Bayesian pathway analysis. There would have to be some hint, or tip, that a particular group might be terrorists to trigger a monitoring of its activities along pathways to CBRN weapons.

Pathways to CBRN

The first step is independent of any information about suspect groups. It is asking: what information is important?[22] What should be collected? This begins, in the CBRN example, by considering all of the possible pathways one could pursue to develop a weapon. The rub, to be sure, is that the numbers of CBRN weapons and paths to them is very large. Biological weapons, for instance, include four groups – aerobic bacteria, anaerobic bacteria, viruses and biological toxins. Intelligence might provide some hints for which particular agents have been of interest to terrorist groups.[23] That might lead to selecting, say, anthrax, botulinum toxin, ricin, plague and smallpox as weapons of first concern along biological weapons pathways.

Enumerate, then evaluate tasks and attributes

The second step in the process is to enumerate all of the possible tasks, materials, equipment, facilities, and personnel along any particular pathway. With those pathways detailed – again, an extensive process – the tasks or attributes can be evaluated for their importance. Indeed, the pathways might be divided in several phases – for instance, concept development, acquisition of materials and equipment, weaponization and deployment. If producing a CBRN weapon along any particular path *depends* on performing a particular task or possessing some particular attribute, it might be rated "critical," something to particularly watch for. A task or attribute might be ranked "important" if it were highly correlated with successfully producing a weapon but had adequate substitutes. A particular task might be critical for one path but not for another. For instance, reducing agents to respirable size might be critical to one pathway to biological weapons but not to another. Plainly, the more pathways that depended critically on some particular task or attribute, the more important that task or attribute.

Evaluate uniqueness

If a task or attribute has few applications other than making a CBRN weapon, it is highly unique; by contrast, if that task or attribute has many applications other than making weapons, it is less unique. For instance, acquiring laboratory equipment is of critical importance in producing biological weapons along any pathway, yet that action ranks low in uniqueness. Possible purchasers of laboratory equipment range from students, to hobbyists to an enormous range of legitimate businesses. So paying much attention to that action in the absence of other indications that a group were potential terrorists would simply be a waste of time. By contrast, acquiring enough fissile material to make a nuclear weapon is very unique. It bespeaks an interest in building such a weapon. Notice that the evaluation of uniqueness does not depend on pathway: a particular task or attribute can be given a single ranking for unique regardless of how many paths to which it applies. This step, like importance, gives intelligence a way to determine which items along a pathway deserve focus. Part of this strategy might also be to collect and analyze combinations of observations that, taken alone, may not raise a flag, but taken jointly are alarming.

Evaluate detectability

The final evaluation of tasks and attributes is detectability. This requires a rough judgment about each task's or attribute's inherent potential for detection. Is there some way it manifests itself so it could be recognized or searched for? The ways to ask that question differ across tasks, materials and equipment, facilities and people. This judgment too can be independent of pathways, without reference to the number to which it applies. In the biological weapons example, while acquiring laboratory equipment is critically important, not only does it rank low in uniqueness, it also ranks low in detectability. In contrast, having the services of a microbiologist might rank only as important, not critical, along most pathways, but it might rank as medium in detectability.

Thus, the three key characteristics of potential observations:

- *Importance:* how relevant a potential observation is to weapon development (can it be easily skipped or is it critical to the pathway)?
- *Uniqueness:* how indicative an observation is of weapon development (are there a lot of other explanations for the observation that are not related to developing a weapon)?
- *Detectability:* how much each item on a weapon pathway lends itself to being observed?

In a Bayesian approach, intelligence can be thought of as an assessment of conditional probabilities. It makes assessments about the likelihood that observations are an indicator of nefarious activities, such as developing a CBRN weapon. The coincidence of observations, any of which might not be alarming if observed alone, taken together might raise the assessment of the likelihood that a person or group is engaging in nefarious activities. Such judgments could then help target further surveillance or other intelligence collection.

For this Bayesian pathway approach, the universe of possible states of the world consists of either: a person or group is developing a weapon *or* a person or group is not developing a weapon. The first state includes cases where the person or group is manufacturing a weapon, in possession of a weapon, considering pursuing a weapon or deploying a weapon for use. Each additional observation made with respect to an individual or group provides additional information – like additional flips of the coin – permitting analysts to sharpen their assessments of which state is most likely for a given person or group. However, in an information-rich environment, it is necessary to have a method to determine what resources should be explored next to maximize the leverage in updating beliefs. Bayesian networks provide a systematic method for updating beliefs that a particular set of observations indicate a person or group is developing a weapon. The approach permits intelligence to identify which observations have the most leverage by exploring how different combinations of evidence impact our assessment of whether a weapon is being developed.

Notice that the CBRN problem is like other transnational "complexities" in that it requires a wide variety of experts – in this case, ranging across the spectrum of weapons and pathways to them, along with people who have some sense for detectability and thus intelligence and other capabilities. Note, too, that detectability can be affected by policy. For instance, in 2007 the US Nuclear Regulatory Commission moved to tighten procedures for getting licenses to acquire radioactive materials after US Government Accountability Office investigators posing as West Virginia businessmen obtained a license in 28 days using nothing more sophisticated than a telephone, a fax machine and a rented post office box.

The three characteristics provide the groundwork for an intelligence framework to detect terrorist weapon development, as suggested by Figure 8.1 below:

Ideally, intelligence would focus on characteristics that were very important to developing a weapon; very indicative of weapon pursuit as opposed to some other benign application; and readily observable. Alas, few potential observables are likely to fall in this portion of the spectrum of characteristics, the portion that is convenient for intelligence. Most will be like acquiring laboratory equipment in the chemical example, if high in importance then low in uniqueness and detectability. Thus, it is shown in Figure 8.1 as high on the importance axis, but near the origin on the other two. By contrast, a microbiologist is less important but shown as more detectable and more unique, though surely not high on either characteristic.

Still, if the objective of a strategy is to detect weapon development, it should concentrate on items on the weapon pathways that rate favorably along each of the dimensions shown – those that fall furthest from the origin. These points are of greatest importance to building a weapon, are fairly unique given a background of other non-weapon activities and lend themselves to observation.

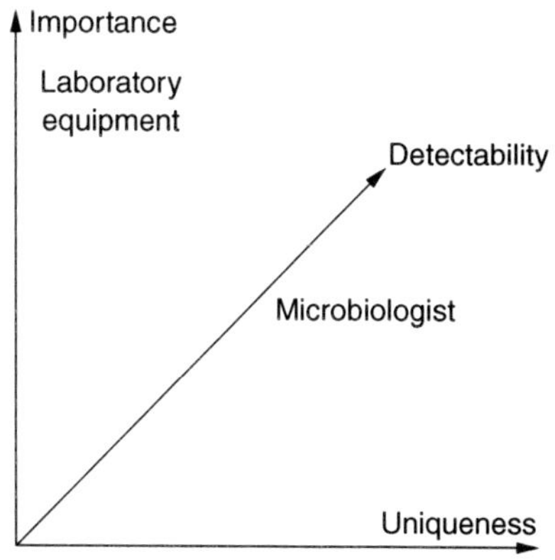

Figure 8.1 Characteristics of intelligence analysis.

Points near the origin represent the area of weakest leverage to address the problem because tasks and attributes that fall here are not essential to completion of the pathway, not unique and not detectable.

For nuclear weapons pathways, plutonium, furnaces and computers are all ranked as being important. However, monitoring purchases of computers is hardly a good component of a strategy to detect nuclear weapon development because computers are not very unique. Plutonium, on the other hand, is quite unique and is also detectable. Spark-free flooring is somewhat unique, but it is not crucial to nuclear weapons pathways and it is only moderately detectable. Note that policies could be implemented to enable spark-free flooring to be detected more easily. For example, the government could require that all purchases of spark-free flooring be recorded in a database accessible to the intelligence community, thus significantly increasing its detectability.

Notice also that uniqueness is context dependent. If the framework were applied not to an industrial society, like the United States or Britain, but instead to a poor region of the world with only a limited industrial base, spark-free flooring might then be considered to be much more unique. Similar considerations may also apply to detectability: transactions that are buried among thousands or millions of others in rich countries – thus obscuring the signal with enormous noise – might stand out more in poorer countries. That might be true whether the activity were observed in surveillance or detected by SIGINT.

The point is then to use the framework as a logic structure for how observations affect the prior belief in whether an individual or group is developing a weapon. The logic develops a set of influence flows between variables.[24] To be sure, the process quickly becomes very complex. Take, as an example, assessing probability that a person or group is developing a radiological dispersal device (RDD). The logic moves from one action to the next along a particular pathway. For instance, if a person were observed investigating sources of radiological materials, that would affect the belief that he or she was interested in acquiring radiological materials, which would in turn affect the belief about whether they were developing a radiological weapon.

Judgments about particular actions can be affected by more than one prior action, thus increasing the complexity. If observations revealed the presence of radiological materials or a Geiger counter, that fact would increase the belief that there is shielding equipment present because these items are often collocated. How much the belief changed is ultimately a subjective matter – unlike the coin toss puzzle – and the judgment will be affected by the context.

For instance, the use or acquisition of a milling machine could indicate that a person or group is pursuing a radiological weapon because several radiological weapon pathways use a milling machine to grind radiological materials into a fine powder that could be dispersed efficiently. However, because milling machines are also common in industry, an observation of only a milling machine would only very slightly increase the belief that a radiological weapon is being developed. By contrast, observing a milling machine *and* shielding equipment might cause much more of an increase. Shielding equipment is more unique, so seeing it alone would increase the belief more than a milling machine alone. But there is synergy between the two, so seeing the two together might increase the belief more than the sum of the two observed separately. If the effects alone were a 3 percent increase in the belief for a milling machine and a 35 percent increase for the shielding equipment, the two together might make a 45 percent increase.

The process provides leverage in updating our beliefs about whether an individual or group is developing a radiological weapon. It could also be used to determine the actions of greatest influence in assessing the likelihood a weapon is being developed – another guide to where intelligence should look and what it should collect. For example, if shielding equipment already has been observed, what action should be focused on next to provide the greatest leverage in determining whether a radiological weapon is being developed? The analysis could also narrow down types of information that might be useful for intelligence agencies to collect (for instance, solicitation of expert opinion versus equipment).

To be sure, the process is complicated, and it lacks the time urgency that is likely to be characteristic of sensemaking with regard to other intelligence challenges of dealing with terrorist complexities. But it does suggest something about the nature of the process that will be required for many transnational issues – Bayesian, iterative, involving a varied set of experts and conducted in the awareness that uncertainty cannot be resolved.

Notes

1 On the distinction between puzzles and mysteries, see Gregory F. Treverton, "Estimating Beyond the Cold War," *Defense Intelligence Journal*, 3, 2 (Fall 1994); and Joseph S. Nye, Jr., "Peering into the Future," *Foreign Affairs*, 77, 4 (1994), pp. 82–93. For a popular version, see Gregory F. Treverton, "Risks and Riddles," *Smithsonian*, June (2007).

2 Dave Snowden, "Complex Acts of Knowing: Paradox and Descriptive Self-Awareness," *Journal of Knowledge Management*, special issue, September (2002), available at: www.kwork.org/Resources/snowden.pdf (accessed December 17, 2003).
3 I spell out this distinction in: Gregory F. Treverton, *Rethinking National Intelligence for an Age of Information* (Cambridge: Cambridge University Press, 2001), pp. 43–6.
4 For a compilation of polls, see AEI Studies in Public Opinion, *America after 9/11: Public Opinion on the War on Terrorism, the War with Iraq, and America's Place in the World*, March 26, 2004, available at: www.aei.org/publications/pubID.16974/pub_detail.asp.
5 See *Country Reports on Terrorism*, released by the State Department Office of the Coordinator for Counterterrorism, April 28, 2006, available at: www.state.gov/s/ct/rls/crt/2005/65970.htm.
6 See Paul Slovic, "Perceptions of Risk," *Science*, 236 (1987), p. 281.
7 The studies suggest that people are much more willing to accept "voluntary" risks, like skiing, than "involuntary" ones, and "controllable" ones than "uncontrollable ones" (Slovic, op. cit., p. 282).
8 James Fallows interviewed a number of terrorism experts five years after September 11. What he found was near-consensus on a high degree of optimism about America's situation with respect to al-Qaeda, "better than many Americans believe, and better than nearly all political rhetoric asserts." See James Fallows, "Declaring Victory," *The Atlantic*, September 1, 2006.
9 For a provocative analysis of the political and social impacts, see Gilman Nils, Peter Schwartz and Doug Randall, *Impacts of Climate Change: A System Vulnerability Approach to Consider the Potential Impacts to 2050 of a Mid-Upper Greenhouse Gas Emissions Scenario* (San Francisco: Global Business Network, 2007), available at: www.gbn.com/ArticleDisplayServlet.srv?aid=39932.
10 The DHS-sponsored fusion centers interviewed by the Congressional Research Service averaged one SCI-cleared person in an average staff of 27. See Todd Masse, John Rollins and Siobhan O'Neil, *Fusion Centers: Issues and Opportunities for Congress* (Washington, DC: Congressional Research Service, 2007), p. 26.
11 James N. Rosenau, "Patterned Chaos in Global Life: Structure and Process in the Two Worlds of World Politics," *International Political Science Review*, 9, 4 (1988), pp. 327–64.
12 For a readable discussion of networked phenomena see Albert-Laszlo Barabasi, *Linked: How Everything is Connected to Everything Else and What It Means*, (New York: Plume, 2003).
13 Testimony in the Oversight Hearing on the Constitutional Limitations on Domestic Surveillance before the Subcommittee on the Constitution, Civil Rights, and Civil Liberties, Committee on the Judiciary, US House of Representatives, June 7, 2007, available at: http://judiciary.house.gov/Oversight.aspx?ID=335; "Legal Authorities Supporting the Activities of the National Security Agency Described by the President," US Department of Justice (2006), available at: www.usdoj.gov/opa/whitepaperonnsalegalauthorities.pdf; John Yoo, "The Terrorist Surveillance Program and the Constitution," *George Mason Law Review*, 14, 3, (2007), available at: www.gmu.edu/departments/law/gmulawreview/issues/14–3/Volume14Issue3.php.
14 These techniques, being developed at RAND and elsewhere, are called robust decision-making (RDM). To make decisions, analysts and policy-makers would look for strategies that seem robust across many scenarios. For intelligence analysis, the aim might be to identify outcomes that seem robust across variables.
15 See Karl Weick, *Sensemaking in Organizations* (London: Sage Publications, 1995).
16 *National Security Strategy of the United States of America* (Washington, DC: 2002), pp. 14–15, available at: www.whitehouse.gov/nsc/nss.pdf.
17 Available at: www.whitehouse.gov/infocus/iraq/iraq_archive.html.
18 Available at: www.whitehouse.gov/news/releases/2002/06/20020601–3.html.

19 It also consumed much of my intellectual attention during the Cold War. See for instance, Gregory F. Treverton, *Nuclear Weapons in Europe*, Adelphi Paper No. 168 (London: International Institute for Strategic Studies, 1981).

20 This episode is discussed in, among other sources, Richard L. Russell, *Sharpening Strategic Intelligence: Why the CIA Gets It Wrong and What Needs to Be Done to Get It Right* (New York: Cambridge University Press, 2007), pp. 47–8.

21 Said at a Department of Defense news briefing, February 12, 2002, available at: www.defenselink.mil/Transcripts/Transcript.aspx?TranscriptID=2636.

22 This approach draws on conversations with and work, as yet unpublished, done by my RAND colleagues, Lynn Davis and David Howell.

23 See, for instance, US Department of Health and Human Services, Center for Disease Control and Prevention, "Bioterrorism Agents/Diseases," (n.d.), available at: www.bt.cdc.gov/agent/agentlist-category.asp.

24 See Judea Pearl, *Probabilistic Reasoning in Intelligent Systems: Networks of Plausible Inference* (San Mateo, CA: Morgan Kaufmann Publishers, 1991).

9 Terrifying landscapes

Understanding motivations of non-state actors to acquire and/or use weapons of mass destruction

Nancy K. Hayden

One horrific September terrorist attack, in the United States, sent the stock market reeling and sparked anti-immigrant sentiment.

Another attack, in Madrid, plunged Spanish Politics into turmoil over issues of war and peace.

Politicians in the U.S. took to describing the war on terror as a struggle of good versus evil.

Religious leaders, quoting scripture, proclaimed that the end of the world was at hand…

… The Year: 1901

Walter Laqueur, *The World of Terror*

Everything that can be counted does not necessarily count; everything that counts cannot necessarily be counted.

Albert Einstein

Background

Concerns about non-state actor acquisition and use of weapons of mass destruction (WMD) against non-combatants have existed since the turn of the century, if not before. However, the level of concern has risen significantly since 9/11, as is reflected in the pervasiveness of the topic in international security policy formulation, national-level political agenda-setting and debate, and emergency preparedness planning. Reasons for the increased concern include:

- widespread perceptions that the events of 9/11 marked the crossing of a threshold in terrorist constraint and lethality;[1]
- open-source accounts of interest in WMD technology by non-state actors;[2,3]
- increased availability of WMD technology;[4]
- greater media attention;[5]
- persistent military presence of the West in global affairs and upsurge of anti-West sentiments;
- sophisticated exploitation of the internet by subversive, non-state actors to build support networks and promote ideology;

- Heightened awareness of the vulnerability of industrialized societies to perturbations in infrastructure operations.[6]

However, having increased concerns about a threat, while it may be wise, is not the same thing as facing an actual increase in a threat and understanding it. A key question is: Have the heightened concerns about WMD since 9/11 been matched by advancements in understanding and characterizing the real nature of the threat based on impartial and holistic use of all relevant information and knowledge? Such an understanding is vital for effective, sound decision-making in national security and foreign policy. As John Parachini (2003) put it,

> although hedging against terrorists exploiting the catastrophic potential of CBRN weapons is an essential task of government resources ... attention cannot simply result in obsessing over CBRN effects but also must produce improved understanding of the motivations, vulnerabilities, capabilities, and context for actual attacks, not just expressions of interest.

Dr. Brad Roberts (2002), of the Institute for Defense Analysis (IDA), noted the paucity of research that exists for understanding the relationship between motivations to acquire WMD on the part of non-state actors, the actual development/acquisition of capabilities and how countermeasures on the part of states affected those motivations. In a conference at Los Alamos National Laboratories, he assessed that,

> experts interested in the terrorism subject had devoted only a tiny fraction of their time and effort to thinking about weapons of mass destruction. Similarly, experts on weapons of mass destruction had devoted little time and effort to thinking about terrorism.

The Center for Nonproliferation Studies, in their 2002 literature review of open-source work on modeling terrorist actions, came to a similar conclusion, stating that "with regards to the specific question of the terrorist decision to employ WMD, the project team could uncover no current research on modeling this aspect of terrorist behavior" (CNS 2002).

Explanatory models of actors' strategic choices involving WMD can potentially draw upon a wealth of research in the behavioral sciences regarding decision-making from a number of perspectives – economics, political and military science, sociology and group dynamics, cultural anthropology, management theory and psychology – and the interdisciplinary developments between these fields. However, Dr. Jeff Goodwin (2006) from New York University recently made the following assessment with respect to scholarly contributions:

> Before 9/11, terrorism research was the exclusive preserve, with very few exceptions, of small networks of political scientists and non-academic "security experts", relatively few of whom were interested in social science

> theory. Descriptive case studies abound, replete with ad hoc, case-specific explanations of terrorism. Curiously, most scholars of rebellion and revolutions have had virtually nothing of significance to say about terrorism. More generally, the strategic choices of social movements – of which terrorism is one – have received much less scholarly attention than the causes and consequences of such movements.

This contrasts sharply with the relevant literature on political leadership decision-making in conflict situations, which draws not only from the field of political science, but also cognitive psychology, small group and organizational dynamics, risk management, sociology, economics, history and cultural anthropology.[7]

Objectives

This chapter presents an overview of recent academic research that addresses motivations of non-state actors to acquire and/or use WMD, and suggests what additional knowledge domains should be considered that have not yet been brought to bear on the problem. The primary (though not exclusive) target was open-source, academic research in the past 5–10 years, accessible to Western scholars (though not exclusively Western in origin).[8] The material was organized as a sort of topological mapping of the "knowledge landscape," addressing the following knowledge contours: (1) intellectual topology; (2) data and research methodologies; (3) common, congruent, and contradictory themes; and (4) research gaps and unanswered questions.

Historically, models of non-state actors' decisions to acquire or not acquire and/or use/not use WMD have been concerned primarily with detection of capabilities and intervention of acquisition. They have accordingly focused on the assessment of technical and operational factors. However, motivations are critical to understanding decisions about WMD – such as why the use of violence and destruction to begin with, the degree of violence and/or mass destruction, legitimacy and instrumentality of targets and timing of use. In addition, motivations strongly influence the degree of effort that an actor is willing to expend to overcome barriers for WMD acquisition, the risk an actor is willing to assume in using WMD, the benefit perceived by an actor in its use, and the feasibility of dissuading an actor from WMD use.

Understanding motivations of non-state actors' WMD acquisition and/or use is complex. Different contexts for different actors must be analyzed in an ever-changing landscape of intentions and capabilities that are constantly being re-shaped by targeted interventions as well as exogenous factors. This chapter makes no attempt at such an analysis. Instead, the purpose is twofold:

1 Review recent research in the behavioral and social sciences for understanding non-state actors' motivations for acquiring WMD.
2 Suggest analytic methods for deeper academic and political understanding and interdisciplinary research to strengthen the underlying knowledge base.

Specific questions guided the reviewing process to make it most relevant for policy-making and action. For example: What does current research tell us about how motivations of non-state actors evolve differently from state actors? What is the role of other internal actors and culture in this evolution? International actors? Is the role of WMD an end point of a "violence spectrum" that may be considered to achieve goals and/or express a message, or is it a fundamentally different choice from other terrorist or insurgency methods? To what degree could the choice to use WMD be dependent upon particular styles of decision-making calculus? What are the influences of contextual and situational factors (apart from culture)? What evidence exists that indicates alternatives to violence or WMD are being considered?

Previous analyses of such questions have been primarily based on case studies of terrorist campaigns and strategic personalities examined through the lens of political science. Using arguments grounded in the rational actor paradigm of decision-making,[9] the aforementioned IDA study concluded that those *least likely* to seek mass casualties were leftists, national and ethnic separatists, state-sponsored groups and cyber criminals; *most likely* were right-wing, transnational and states pursuing asymmetric strategies in war against the United States. Those most likely to use WMD were judged to be additionally motivated by religious ideologies. In summarizing this work, Roberts reported that literally no experts had written on transnational terrorist use of WMD (Roberts 2002).

What, if anything, has changed since these reports were issued?

Framing the problem

Defining weapons of mass destruction

Studying motivations of non-state actors for WMD acquisition is complicated by the lack of consensus on what constitutes a weapon of mass destruction. According to Robert Whealey, the term was first used in 1937 by the *London Times* to describe the German aerial bombardment of Guernica in Spain. This attack, ordered by President Franco to crush the Basque resistance to Nationalist forces, destroyed 70 percent of the town and killed one-third of the population. The phrase, "weapons of mass destruction" was coined to describe the massive amount of damage caused by conventional bombs in this new military tactic, aimed against a civilian population with the intent of demoralizing an enemy (Mallon 2008).

Since that time, the term WMD has come to have different meanings in different contexts that range from international treaties and courts of law to Wikipedia. The vernacular definition in Wikipedia, based on a synthesis from a number of reference sources that include encyclopedias and science dictionaries, is:

> Weapons of Mass Destruction is a term used to describe munitions with the capacity to indiscriminately kill large numbers of living beings. The phrase

> broadly encompasses several areas of weapon synthesis, including nuclear, biological, chemical (NBC) and, increasingly, radiological weapons.[10]

Here, the meaning is focused on the weapon as a means of – and potential for – destruction, not the destruction itself. This is an important distinction, with implications for studying motivations. In national policy circles, the term often has an even more narrow meaning that is limited to specific technologies. Based on terminology in many official US government documents, the Center for Nonproliferation Studies (CNS) at the Monterey Institute of International Studies defines "weapons of mass destruction" simply as "nuclear, chemical, and biological weapons."[11] The CNS notes, however, that those US laws and documents which are focused on responding to possible WMD incidents in the United States often take a broader view of WMD, and include in their definition radiological weapons or conventional weapons causing mass casualties.

In a statement to the Emerging Threats and Capabilities Subcommittee of Armed Services Committee of the United States Senate, Major General Robert P. Bongiovi, then acting director of the Defense Threat Reduction Agency, presents the latter view, testifying that

> The [WMD] definition encompasses nuclear, chemical, and biological weapons. However, it also includes radiological, electromagnetic pulse, and other advanced or unusual weapons capable of inflicting mass casualties or widespread destruction. In addition, conventional high explosive devices, such as those used in the attacks on Khobar Towers and the USS COLE, are legally and operationally considered to be WMD.[12]

Definitions for WMD that focus exclusively on weapon agents are problematic as points of reference for studying the knowledge domains underlying non-state actor motivations for WMD threats. Social psychology literature points out that, in terms of motivations, the important decision threshold to be crossed is whether or not to engage in mass destruction involving the killing of "innocent" noncombatants. (Note: the term "innocent" itself is a loaded term, which depends on one's perspective and world view.) Once that decision has been made, the choice of technology – whether it be conventional explosives used on a massive scale, or something more "exotic" such as chemical, biological or nuclear weapons – is a much smaller step and is often (although not always) driven by factors involving opportunity and expertise rather than social and behavioral considerations.[13]

Therefore, the following definition for WMD used for this chapter does not depend on technology, but rather on the intended effects of the use of the technology its targets: "A weapon of mass destruction is the means or capacity to intentionally and indiscriminately kill or put at risk the well being and livelihood of large numbers of (non-combatant) living beings."

This definition includes, but is not limited to, nuclear, chemical and biological weapons; it also includes cases of destruction of critical infrastructures and other essential resources for life, as well as direct loss of human life.

Defining motivation

The second aspect to delineate is the nature of motivation itself and how it will be considered. Motivation has been studied from many perspectives – e.g. psychological, organizational, educational, biological, political and spiritual. There is no single definition of motivation across these disciplines, nor is there a grand canonical theory to explain it, beyond the most basic principles.[14] Moreover, as Allison and Zelikow have pointed out in their classic analysis of the Cuban Missile Crisis (Allison and Zelikow 1999), political leadership decision-making has complex motivations based on many factors that include the process of and participants in the decision-making, as well as personal and contextual considerations.

In spite of the various views of motivation, there is consistency in conceptual definitions across different scientific domains that provide a coherent basis for this study. The following definition of motivation, is drawn from organizational management literature, and incorporates those consistent concepts: "the forces either within or external to a person *or group* that arouse enthusiasm and persistence to pursue a certain course of action" (Daft 1997).

There are several reasons this definition was chosen. First, it recognizes that the forces which activate behavior can be either internal or external. Second, it is neutral with respect to the nature and origin of those forces. Third, it highlights the importance of persistence. Lastly, it suggests both direction and goal-orientation in the pursuit of action without constraining the definition to rational actor approaches.

Fundamental to the study of motivation for acquisition of WMD are the universal premises derived from Maslow's hierarchy of needs, which bear directly on the correlations between motivation, deterrence and evolution in shaping behaviors:

1 Human needs are either of an attraction/desire nature or of an avoidance nature.
2 Because humans are "wanting" beings, when one desire is satisfied, another desire will take its place.

The first principle was reflected in many of the cost–benefit approaches in the literature, as will be noted below, whether or not the principle was explicitly acknowledged. In general, however, this first principle was underdeveloped, while the second was ignored. Three additional behavior principles related to motivation are important to this research:

3 Behavior is always the result of a combination of drivers.
4 Cognitive beliefs about what will happen as a result of behavior are powerful motivators irrespective of their grounding in reality.
5 The persistence of motivation depends on feedback from the behavioral action taken.

The questions of what constrains non-state actors and how those constraints are shaped by competing needs, internally derived belief systems and feedback from internal and external actors are key to understanding the motivations to acquire WMD, intentions to use WMD if acquired, and the drivers that do or do not come into play. Deterrence theory was developed to explain state-level constraints from the perspectives of political and military science, economics and international relations, alongside the development of international norms. With non-state actors, while some of these same drivers may be relevant, cultural constraints, societal norms and religion become increasingly important. In some cases, they can dominate the analysis. With respect to religious ideologies and motivation, Bruce Hoffman wrote, in 1997 that "Terrorism motivated by extreme interpretations of religious doctrine assumes a transcendental dimension, and its perpetrators are consequently unconstrained by the political, moral, or practical constraints that may affect other terrorists" (Hoffman 1997). Finally, the relationships between all relevant actors is critical to consider in studying motivations for non-state acquisition of WMD, as they provide the feedback dynamics that shape persistence. What factors play into the relationship between religiously motivated extremists and the broader pool of like-minded believers? Are there ideological constraint mechanisms? What role does ideological competition with other groups play? Figure 9.1 shows the multiple directions that these forces project within like-minded groups and between those groups and others in competition for the same resource base of popular support.

These dynamics underscore the interactive nature between the belief systems and behaviors of the multiple actors involved in studying motivation for WMD proliferation. Indeed, the motivations can be viewed as emergent phenomenon of a complex system comprised of these multiple actors – both human and technological. Interactions continuously generate adaptive responses, behaviors

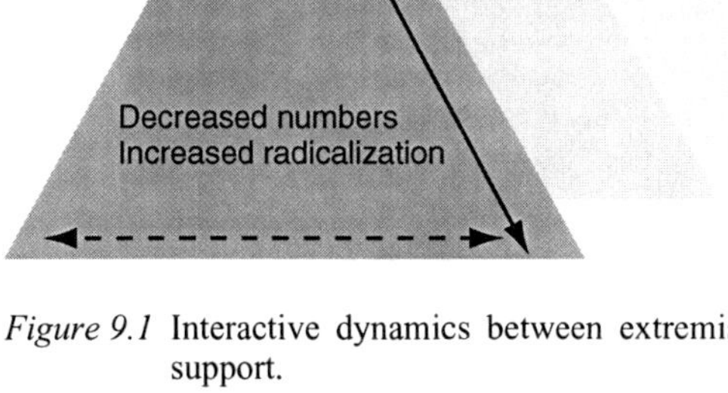

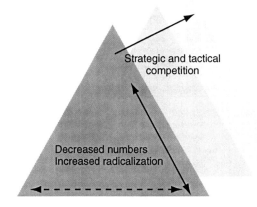

Figure 9.1 Interactive dynamics between extremists and their underlying base of support.

and beliefs. Figure 9.2 shows the multiple feedback paths between the technical capabilities acquired by actors, their motivations, the vulnerabilities of the systems they target, their perceptions of the consequences on the intended audiences, the ability of the intelligence community to detect and interdict the development and/or deployment capabilities and the decisions of policy-makers for intervention (Hayden 2007).

Ideally, a solid research base would provide the knowledge necessary to inform the dynamics between all parts of this "system of systems" at the appropriate levels of granularity. In reality, the research fields and analysis methods don't often tend to overlap; when they do, there is a problem with data methods and semantics. Even within domains, multiple levels of analytic consideration are rarely spanned (i.e. individual, group, organizational and institutional) with consideration of interdependent factors and feedback between layers. As a result, analyses often disaggregate the problem space into categorical actors, viewed through single perspective in a "once-through" mode, rather than considering the problem holistically, as an interacting and evolving "system of systems" seen through multiple perspectives. These multiple perspectives are extremely important when trying to explain the motivations of actors in diverse cultures.

This study undertook to identify research and analysis that support interdisciplinary models, diverse world views, and include domains not normally found together. A conceptual model of those domains and the types of inter-related behavioral factors that they represent, relative to motivations, is shown in Figure 9.3.

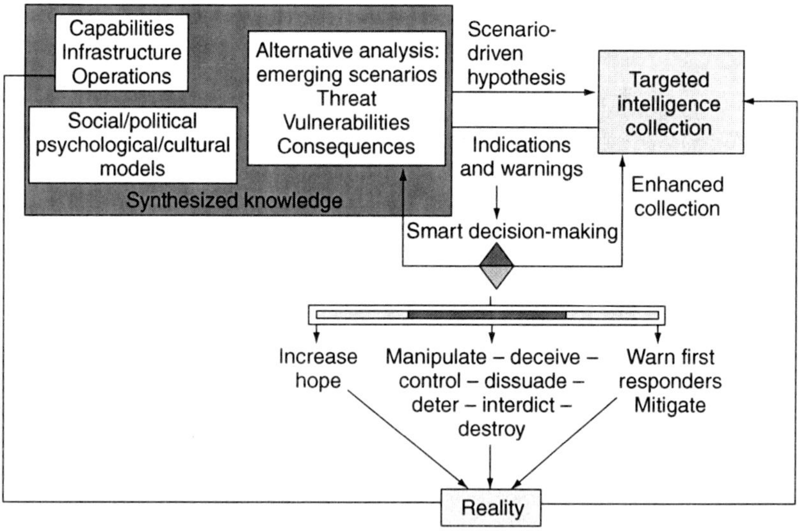

Figure 9.2 Motivations to acquire WMD emerge from complex interactions between multiple actors over time.

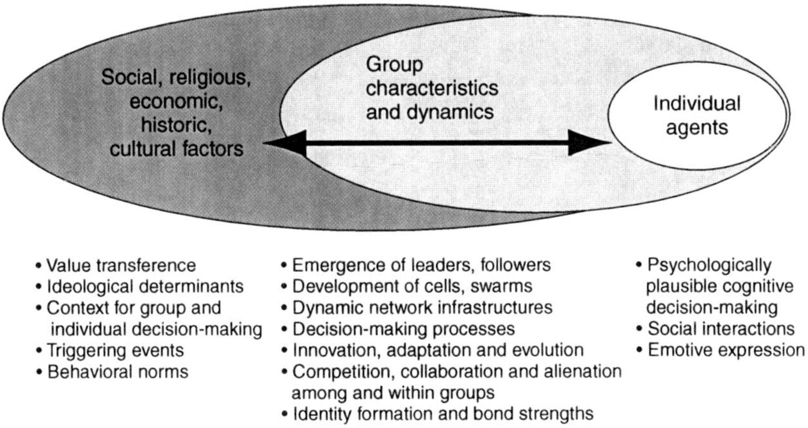

Figure 9.3 Multiple levels of behavioral factors for motivations to acquire WMD.

Wicked problems

The dimensions of, and complexity inherent in the study of motivation described above constitute characteristics of what has been termed a "wicked problem" (Rittel and Webber 1973). This, in turn, has implications for the underlying knowledge base for motivations that must be incorporated into models for WMD proliferation. In the end, it may need to be as diverse, dynamic, flexible and adaptable (i.e. "wicked") as the potential solution space itself. One reason for this is that social, not technical, complexity drives the non-linearities in, and unpredictability of, wicked problems.

A generalization of wicked problem characteristics is that:

1 You don't understand the problem until you have developed a solution.
2 Wicked problems have no stopping rule.
3 Solutions to wicked problems are not right or wrong.
4 Every wicked problem is essentially unique and novel.
5 Every solution to a wicked problem is a "one-shot operation."
6 Wicked problems have no given alternative solutions.

In short, there is no well-circumscribed boundary around the knowledge domains that must be considered for complete understanding of wicked problems. The implication is that, in theory, every consideration of a particular potential non-state actor pursuit of WMD must be able to draw upon a "dynamic" body of literature that is as flexible as the potential motivational responses to interactions between the potential proliferant and other actors in the system. With this caveat – which implies that consideration of the questions posed by this research can never be complete or finished – a corpus of research from a broad (though still somewhat limited) spectrum of social sciences was developed and analyzed. Of particular interest was whether – and if so, how – literature on transnational groups, social movements, extremism and religious ideology, violence and non-violent conflict might inform the question of non-state actors' motivations with respect to WMD over time.

Approach

The data-mining method used was the "snow-ball" approach, starting with the author's own research, previous surveys, institutional resources[15] and publicly available bibliographies and data bases (such as those maintained by the National Defense University, the US Military Academy, the Terrorism Research Center and the Homeland Security Institute). Academic publications, primarily within the past five years, from both the US and non-US perspectives were sought, although the preponderance of material found was Western. Data-driven literature with clearly articulated research methodologies and/or unique data (such as statements by jihadists themselves) was preferentially considered over opinion pieces or political commentary.[16] This led to articles and research in numerous peer-reviewed academic journals, institutional publications and special access and compilation services such as *Jane's Intelligence Review*, the Open Source Center and the Congressional Research Service.

The search was expanded to citations further back than five years, to include important concepts from social sciences that had been overlooked in previous studies. The literature from desired citations were compiled and analyzed according to source, research context and content (see Figure 9.4). A structured database was created to experiment with semantically based natural-language processing, pattern recognition and large graph analysis software tools to discover visually analyzed trends and correlations between concepts, authors and institutions, and to aid in conducting additional web-based literature searches.[17]

Results of those exercises showed potential value in identifying clusters within the citations, conducting time-series analyses of research trends, discovering latent connections between authors, institutions and concept, and making side-by-side comparisons of concepts between different collections of writings. However, one lesson learned early on is that the use of these tools entails a significant up-front investment of data manipulation and programming to get information into a common format that is "clean" with respect to the tool.

Findings

Over 275 relevant citations were collected, with the majority being published within the past six years. Of these, approximately 50 were peer-reviewed journal articles, although only a handful of these addressed WMD directly.[18] A primary goal for this research was to expand the scientific base to include contributions

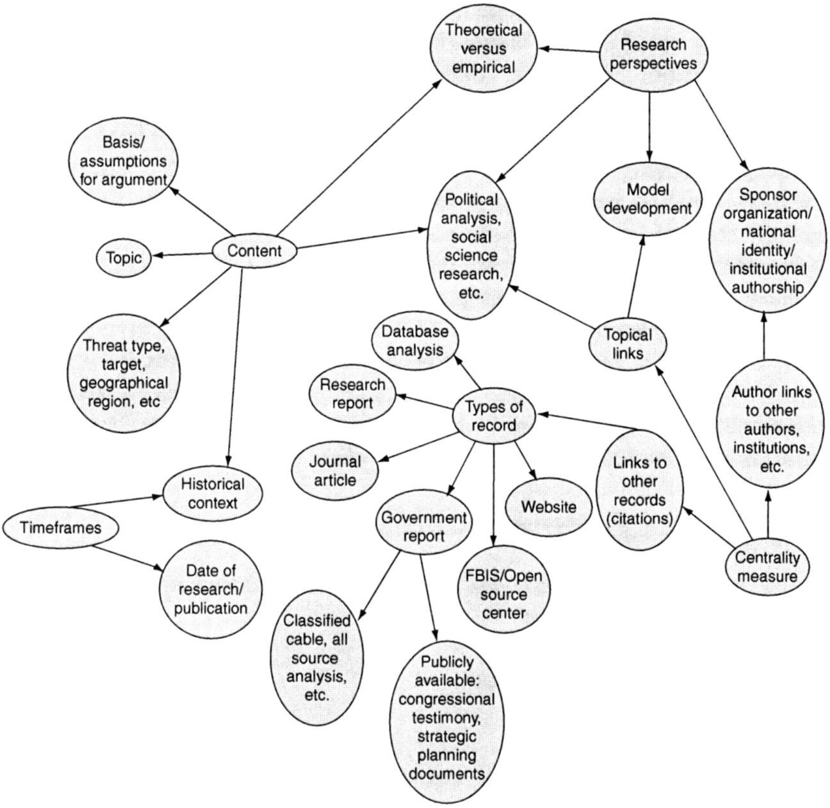

Figure 9.4 Conceptual map of literature sources.

from a broader spectrum of social science disciplines and generate new conceptual models. Accordingly, where they were deemed to have relevance to motivations that lead to the acquisition of WMD, citations were also drawn from literature on group violence; cultural norms on social justice and violence, terrorism and WMD; religion and violence; and apocalyptic world views. Due to resource limitations, the breadth and depth of possible domains searched had to be limited. Future research could, and should, continue to expand this base.

The majority of studies that directly address the WMD motivations came not from peer-reviewed social science literature, but from reports of think-tanks, security and policy research institutions or thesis papers. The thesis papers were primarily from US military academies, and drew heavily upon political science traditions and historic analogues. For the most part, this body of WMD literature was based on case studies of the very limited instances where WMD have been used, or an attempt to use them has been made, with the definition of WMD being limited to chemical, nuclear or biological weapons.

One of the most interesting areas of consensus in this literature is that, based on analyses of the limited case studies available, WMD use by non-state actors has proven to be relatively *ineffective* in causing mass casualties (Parachini 2001). That historic precedence, combined with arguments about the difficulties in acquiring and using WMD and the political and social/cultural constraints against indiscriminate mass casualties in general, form the basis of fairly uniform arguments about the primary "de-motivating" factors for WMD use. The political constraints draw on rational choice theory, strongly coupled with assumptions about (1) the role of social/cultural norms and their importance in maintaining a sympathetic base of support from the populace and/or (2) the state responses perceived to be likely. These assumptions have not been proved empirically, however, and are identified as being in need of further research.

So, what are the motivational factors that might entice actors to overcome these obstacles and pursue WMD? Which actors are most likely to have these motivations? What does current literature – especially that since 9/11 – have to say, in comparison with the "conventional wisdom"?

Conventional wisdom

Previous research widely agreed that motives for WMD acquisition by non-state actors were mostly likely to be:

- attract more attention to a cause
- create economic havoc
- hasten the apocalypse
- promote a worldwide race (culture) war to establish a homogenous state
- create an aura of divine retribution
- impress target audiences with high technology
- copy-cat tactics.[19]

Regarding the question of who is most likely to act on these motives, a fairly consistent causal story had emerged based on three key hypotheses:

1. motivations of non-state actors are different than state actors with corresponding differences on constraints;
2. non-state actors most likely to use WMD have apocalyptic/millennial beliefs with a sense of a global mission; and
3. the nature of a "new" terrorism (characterized by a transnational and networked nature of actors) facilitates the amplification of motivations and support among sympathizers, thereby reducing constraints otherwise provided by norms in face-to-face groups as well as at the social and cultural levels.

These premises support a tightly linked argument prevalent in much of the literature for assessing WMD motivations that has become "conventional wisdom," as noted by Adam Dolnik (2003). Namely, that the primary threat of WMD by non-state actors, both in terms of use as well as acquisition, is from transnational, apocalyptic and religious groups. However, as noted by Gary Ackerman commenting on his own survey of books, journal articles, monographs and government reports on WMD terrorism,

> the scholarly and policy-related literature has increasingly begun to recycle the same interpretations and staid shibboleths. This is not meant to denigrate several excellent works that have emerged, merely to point out that truly novel insights into WMD terrorism are becoming few and far between.
>
> (Ackerman 2005)

An objective of the research for this chapter was to discover to what extent the conventional wisdom regarding motivations was still prevalent, to what extent was it being challenged by new evidentiary-based research and/or conceptual approaches, and how the evidence for this position compares to that for alternate viewpoints. A key finding is that, to a large degree, the conventional wisdom continues to dominate the literature. Exceptions are in three areas: research that examines the nexus of transnational criminal organizations, terrorist groups and state institutions (or the lack there-of, such as in ungoverned spaces); research whose focus is on leaderless jihadist groups; and research into the evolving role of the internet in promoting global incentives through online cliques, groups and communities and the ease with which technology transfer is facilitated into such groups. There is also a proliferation of literature purporting to explain whether or not Islam, as a religious ideology, provides either a motivation or a constraint (with both sides being argued).

There are several explanations for the persistence of the "conventional wisdom" about WMD motivations. One is that, while the original logical arguments may have been robust, there has been little ongoing re-evaluation of data to ascertain their validity in evolving contexts. As a result, there have been relatively few new hypotheses put forward to test ideas about traditionally characterized non-state actors. Another is the relatively small domain of subject matter expertise that has been tapped in coming to these ideas – as noted by Dr. Goodwin's remarks cited earlier. Yet another is the limited empirical base that has been used to develop and test hypotheses; to date this has rested primarily on the few case studies of WMD use that exist, alongside trend analyses of past terrorist events. This, in effect, tests only on the behavior of interest (use of threat of mass destruction by non-state actors to achieve end goals), and does not provide adequate control groups for developing robust causal models for decisions to use or not use WMD by the entire spectrum of potential non-state actors.[20] A fourth explanation is that much of the literature is produced within narrowly defined fields of study (e.g. psychology, military operations, political science, sociology, cultural anthropology, etc.) without cross-over between fields through interdisciplinary approaches (and by extension, without considering interaction affects or correlations between behavioral data across disciplines). A fifth reason is that the work of Western scholars dominates the field, with all of the attendant assumptions, world view and value systems that implies. Finally, as noted in several recent papers, the causal factors for motivations to acquire and/or use WMD by non-state actors has been treated as a deterministic, first order problem when in reality, it is a "wicked problem," driven by second order, non-linear affects that are only just beginning to be studied.

New looks at data

Comprehensive structured data sets with internal consistency that can be used to deduce the behavioral motivations underlying terrorist incidents has suffered from a multitude of problems, as several researchers have noted (Atran 2006). One of the primary criticisms is that these data sets have focused narrowly on details of the events (i.e. methods used, casualties, fatalities, targets, perpetrating organization), isolating them from information about the context of the event, responses to the event and subsequent activities of terrorists, and behavioral characteristics of the perpetrators of the event. Other limitations to developing globally relevant causal behavior models are the lack of consistency across different data sets gathered at different times and covering different geographic regions, and the attribution of significantly different definitions to the events characterized, resulting in different event counting methods.

With the increased interest in the behavioral drivers behind terrorism in general, and questions about potential WMD use specifically, there has been an increase since 2001 in the amount and types of data gathered on terrorists and their networks of support, and terrorist incidents in general that are just starting to become available for the broad research community. One such database is the Global Terrorism Database (GTD), which covers over 69,000 incidents since 1970 – including religious, economic and social acts of terrorism, as well as political (LaFree *et al.* 2005). Sponsored by the Department of Homeland Security and maintained by the National Center for the Study of Terrorism and Responses to Terrorism (START) at the University of Maryland, the GTD combines global terrorist incident data collected by Pinkerton since the early 1970s with data compiled by RAND Corporation. Further work on the GTD has been done to redress problems encountered in other databases around issues of consistency in counting methods, comprehensiveness of geopolitical coverage and breadth of characteristics captured.

Analysis of the GTD in its beta phase has shown trends that challenge the conventional wisdom. Namely, that "the willingness of traditional terrorist groups to engage in mass casualty terror indicates that they may be more likely to acquire and use WMD than has been assumed by the majority of analysts" (LaFree *et al.* 2005).

Future studies of this data might allow more detailed analysis around questions such as, "Under what conditions have groups who have shown a willingness to use mass violence been deterred from doing so again?" and "Under what conditions have groups who have shown a reluctance to use mass violence changed their strategies?"[21]

Only a handful of other researchers have posed similar questions about motivations for WMD use in the expanded context of overall propensity to engage in mass casualty terror. Those who have, such as Parachini (2001), Dolnik (2003) and Asal and Blum (Asal and Blum 2005), find results similar to LaFree. Dolnik reviewed the statistical inferences of Bruce Hoffman in 1998 using the St. Andrews–RAND database. Based on potential bias and errors in the reporting of the data, and his own analysis, he questioned Hoffman's conclusion that rises in lethality of terrorist incidents could be attributed to Islamic extremism. Using the Incidents of Mass Casualty Terrorism database, created by Robert Johnston,[22] Asal and Blum argue that: (1) contrary to the conventional wisdom, groups engaging in mass casualty attacks to date have achieved symbolic impact through their choice of targets, not their choice of weapons; and that (2) while a clear plurality of mass casualty terrorist events have been perpetrated by terrorists from the Islamic world targeting the non-Islamic world, terrorist attacks have run the gamut of religious and ethno-nationalist motivations.

These data-based analyses bring new insights for the broader question of using motivations for mass casualty terror in general as a precursor to understanding WMD motivations in particular. However, they have not yet become widely cited or replicated in the literature.

At the meta-level, one can observe that collections of literature on WMD and non-state actors (terrorism), collected and communicated via the internet as bibliographies, special projects or online resource libraries are proliferating. Indeed, the number of such sites far exceeds (by at least an order of magnitude) the sites with data. Many of these are generated and maintained by the same communities that are active in generating the literature itself – such as the defense community, political think-tanks, political activists and the national security community.[23] As a result, there is a core set of literature and authors that appear frequently. A thorough comparative analysis of these sites was not done. However, qualitatively it appeared that the sites themselves shared the same characteristic as the literature; that is, there is much self-referencing and repetition of ideas and authors across the majority. With a few exceptions, while the corpus contained within each may contain excellent material for familiarizing students and newcomers to the topic with basic concepts, the lack of alternate perspectives and hypotheses, the absence of interdisciplinary approaches, and the lack of interactive dialogue and debate evidenced in the literature limits the intellectual value for deepening understanding among seasoned experts, analysts and policy-makers.

New ideas

The question of which non-state actors are most likely to be motivated to acquire and/or use WMD has traditionally been analyzed according to some variation of the basic psychology-based typology of terrorist groups proposed by Jerrold Post, in which groups are differentiated by their purported raison-d'être; namely, social revolutionary, nationalist-separatist, religious fundamentalist, religious extremists (closed cults), right-wing extremists and single-issue extremists (Post 2000).

These categories continue to be used, with two notable additions since 2001 – the increasing attention being paid to linkages with transnational criminal groups and to network-based terrorist groups. Dr. Louise Shelly of the American University has found that, whereas traditional organized crime entities prefer stable and predictable institutional structures for bases of operations, a different type of transnational criminal group element is flourishing in post-conflict situations. These groups thrive on chaos and tend to form partnerships of convenience with terrorists in separatist areas and ungoverned spaces, with motivations that differ from the traditional organized crime models (Shelly 2007). Another new category of non-state actor for WMD proliferation is the network-based, leaderless jihad groups. The motivations of these groups are hypothesized by Marc Sageman to be moral outrage, which in time may become self-terminating (Sageman 2007).

Analysis of high-stakes decision-making differs, depending on whether it is viewed through rational choice models grounded in political science paradigms, versus those based on sacred values and drawing on the field of cultural anthropology.[24] In the case of the former, Stohl is representative of those who evoke expected utility theory. He has developed a framework for assessing the strategic value of WMD through this lens, based on six questions around capabilities, motivations and resources (Stohl 2005). Using a similar paradigm, Lichbach demonstrates that contradictory conclusions can be reached even through rational choice theory, depending on whether one takes the view of the bargaining theory of war or the view of the rebel's dilemma theory of dissent to support acquisition and perhaps even limited use of WMD (Lichbach 2003). On this basis, he argues that histories of conflict impact motivations to acquire WMD – depending on whether those histories encourage or discourage the involved parties' abilities to forge credible commitments. These examples show more complex uses of rational choice theory being employed that account for an increasing number of factors and options.

Anthropologists are beginning to make their appearance in this literature, arguing for sacred values as a motivational power for extreme behaviors – especially those which call for sacrifice of individual lives to provide meaning for those lives.[25] Here, there is a tension between levels of needs in Maslow's hierarchy: that of self-actualization (sacred values) and those of safety, love and self-esteem (rational choice). The degree to which sacred values trump the rational choice model of maximization of utility and are prime drivers for extreme behaviors (versus an exploited tool of justification of otherwise-motivated behavior) remains a debate among scholars. To complicate the picture, Asal and Blum have taken a new look at data on mass casualties and their relation to sacred values with respect to potential WMD acquisition and/or use, and have demonstrated that the data on mass-casualty attacks does not support the conclusion that apocalyptic, religious groups are the most likely to engage in mass-casualty attacks. On the basis of data in the period 1970–2002, they argue that groups with strictly religious motivations account for less (23 percent) of the attacks than those with ethno-nationalist causes (32 percent); while a substantial number (14 percent) are ethno-nationalist based, framed within a religious context (Asal and Blum 2005). This comparative study does not negate the role of sacred values, but does call into question the degree to which they can be considered explanatory absent other factors.

To deconstruct the complexities of the motivations in these cases requires more comparative studies of the convergent roles of religion, ethnic identity and social grievance in the WMD context. Such studies are starting to appear in the WMD literature – most notably the work of Asal and Blum, Ackerman, Atran, and Sagemen. However, this work tends to be done either at the group–group comparison level, or at the individual level. Much more needs to be done, controlling for interaction factors between group-level decisions, individual motivations and the social/political context.

New research on societal level factors that impact WMD motivations of non-state actors include the "socio-autoimmune" effect proposed by Steinbruner (in which over-reactive responses by the attacked society are strategically exploited) (Steinbruner 2005); the role of diaspora communities (Sheffer 2005); the risk posed within minority communities (Ramraj 2006); the previously mentioned role of transnational organizations and international criminal networks;[26] the impact of specific cultural geopolitics (Lacroix 2004; Tan 2003); generational effects (Drennan 2006); and the relationship between global conditions, such as climate and environmental degradation, and mass murder (Leven 2002). The influence of global trends on factors important to motivations for WMD acquisition has not been studied to date. The implications for potential strategic use of WMD by some of the "more pragmatic-minded" non-state actors are just beginning to appear in research literature.[27]

A new development during the past five years is the abundance of publicly disseminated and debated extremist actors' statements about WMD. These have become a rich resource for analysis of strategic and tactical intent.[28] While this is not a new trend in general terrorism research, here-to-fore there was little use made of such primary material in the nexus between non-state actor WMD research and terrorism studies. The internet increasingly provides an unprecedented accessibility to, and transparency of, internal and external debates among these actors. This is particularly true with respect to Islamic and right-wing militia groups using the web for dialogue and communication among supporters and for recruiting. The internet resources are supplemented by unmediated "publicity campaigns" for extremist causes that show more and more sophistication, exploiting the broadcasting industry, print and digital formats. These are accompanied by self-promotion in the forms of interviews and video productions.

While providing potentially valuable resources for analyzing intent, the way in which statements by terrorists are used by Western scholars and analysts must be considered carefully. Western scholars untrained in the Arab language and/or Islamic culture will have a limited ability to understand nuances carried by certain images and phrases. Another issue is the inherent difficulty in distinguishing the ultimate intent of the statements – when can they be taken at face value and when are there underlying motivations and contexts that have been obscured – either intentionally or by happenstance? The specific audience for which the statements were intended must be considered in order to understand the communication. There is also a question of the degree of persistent influence that is wielded by these statements upon the intended audiences. Finally, with respect to prison interviews and "confessions," there are well documented issues in research literature regarding the caveats that must be placed on such data, due to the tendency of some individuals in these circumstances to say what they believe they are expected to say, or what is to their best advantage to say.

With these caveats in mind, collections of statements were analyzed for their potential to inform analyses of WMD motivations. The most complete translated compendiums are those done by the US Military Academy as part of the Harmony Project. The Harmony Project was started in 2002 to translate and analyze al-Qaeda documents for strategic content. Of primary interest to motivations, is the report on the military ideology of al-Qaeda by McCants and Brachman (2006).[29]

It is striking that, in the relatively massive compilation of strategic documents collected, WMD plays a very minor role in the overall scheme of militant ideology and strategy of al-Qaeda. While contrary to perceptions of al-Qaeda's purported interest in WMD, it is consistent with observations of scholars that, in spite of rhetoric to the contrary, the weapons of choice for mass casualties by Islamic terrorists to date have been conventional, as opposed to WMD, technologies. When WMD is mentioned, it is usually endorsed on the basis of the overall principles of jihad. This obscures what the real motivations might be, which, as pointed out by Parachini in 2001, can range from fascination with technology to victimization narratives to act of cosmic war, and many in between. In general then, these sources, while they have a rich potential, currently portray a confused and ambiguous picture regarding WMD motivations.

New contributions from expanded domains of expertise

There are some scholars who are confident that, based on evidence of "interest in" WMD, there is no doubt today as to which groups seek to acquire WMD and those which do not.[30] However, even should the evidence be deemed appropriate for inferring intent, it still does not explain motivation and the inter-relationship between the groups themselves and their base of support and how that affects motivation. A recurrent theme throughout WMD literature is the need for better understanding of the causes of extremism and mass violence; the complex, multi-actor dynamics of collective action in contentious politics; and the role of culture, historical narratives, social institutions and identity formation in shaping those dynamics. Such knowledge is essential for understanding early warnings and developing effective deterrence policies. Literature from the fields of social psychology, history, religious studies, organizational and group dynamics, communications and anthropology contain rich material for addressing these issues which as yet are underutilized in exploring the question of WMD.

Working groups and position papers on radicalization and de-radicalization processes have proliferated in the past several years, along with statements of its seriousness as a threat by various government organizations. Very few scholarly research papers have been generated, however. Of those that have, there is a diversity of paradigms employed, including social movement theory (Bell 2003), rational actor models (Ferraro 2005), in-group dynamics[31] and alienated clique formation.[32] What is not clear is when, why and how each of these models might be operational and what the observable markers are for the existence of the process.

Another key question relating radicalization to WMD is whether or not these models suggest that actors move along trajectories toward mass violence to advance extreme ideologies, or whether there is a discontinuity in the types of choices made. There is little research done on the non-violent side of radicalization where alternative means are chosen to advance extremist ideologies.

Sociologists' research on collective action and violence should be mined further to explore questions of causal mechanisms that move individuals and groups from belief to action, from emotion to action and combinations of the two.[33] Communications theories, social networks and influence dynamics also provide perspectives for studying the spread of ideas, and psychological and cognitive decision-making theories provide understanding for how these ideas cause movement from belief to action. These knowledge domains were found to be used in studying violence, but not specifically cited in WMD motivation research. Religious violence, especially with respect to political Islam, emerged in this study as a category in itself that warrants deeper comparative analysis than has been done to date for looking at conditions under which mass violence has been motivated by sacred values, and under which the same sacred values have channeled behavior away from indiscriminant mass violence. A final research category emerged around the framing of the problem – dealing with parameterization of concepts, formation of modeling and simulation tools, and risk assessment methodologies.

Non-linear and second order affects

As noted earlier, the motivation to acquire WMD by non-state actors is complex, and occurs in a dynamic and evolving context. This research perspective is beginning to appear in the most recent literature, where the need to study non-linear and second order affects has been specifically called out, especially when making judgments about trajectories of future developments.[34]

The debate in the jihadist strategic literature over the use of WMD illustrates one aspect of this. In 2003, the Saudi cleric Nasir Bin Hamd Al-Fahd, citing the damage of American bombs in Muslim lands, used the pronouncements of historic Islamic jurists to legitimize the use of WMD. On the other hand, Tariq Ramadan, who purportedly shares similar ideological visions, publicly exhorts Muslims against terrorism and violence to achieve them. Who holds more sway among audiences, which audiences are most likely to take action, and what is the impact of the contradictory views of these two leaders? These dynamics, arising out of ideological competition among leaders for followers, are not well understood, especially by Western scholars, within the Islamic context.

Other unresolved questions that surface repeatedly in the literature that bring into play second order affects revolve around the interaction between motivations and technical capabilities, perceived vulnerabilities, and impact of state actions. How do repressive versus facilitative state responses impact the motivations of followers? What is the impact of surveillance on the crystallization of extremist beliefs within groups? How do transnational trends in terrorism organizations – such as connection to criminal networks – change the motivational factors and constraints?

One can begin to build a two-dimensional decision space to map the intersection of the most important of these, with endogenous factors on one axis and exogenous factors on another. The research for this chapter suggests that one dominant endogenous variable – common to virtually all of the literature – may be the degree of internal group cohesion that is present and necessary to sustain the non-state actor in accordance with the group's goals and operational modes; while a dominant exogenous factor may be the strength and direction of relationships of the actor to external institutions (political, social or religious) necessary for its power base.

Conclusions

An extensive (though not exhaustive) collection of scientific publications was generated and analyzed to a first order, drawing on primarily Western-based academic publications over the past five years, which purport to examine the motivations of non-state actors to acquire and/or use of WMD. The overall landscape of this literature has changed somewhat, though not dramatically, since 2000. The following trends, gaps and outliers were noted in particular:

1 Capabilities versus motivations: with respect to WMD research, the majority of Western-based work continues to focus on capabilities of non-state actors, rather than on motivations.
2 Much (though not all) of the research on motivations of non-state actors to acquire WMD is generated from the political science discipline (not behavioral sciences) and draws heavily on relatively static ideas of rational actor theory, bargaining theory of war, or the rebel's dilemma with its asymmetric power relationship. The second most dominant discipline to contribute to the research literature on motivations to acquire WMD by non-state actors is the field of psychology.

   These ideas have not changed much in the past five years, and many of the "old questions" around the interfaces between states and non-state actors that existed in 2000 remain largely unaddressed by data-based research. Some of these are: With respect to acquisition of WMD, do state and non-state actors influence and/or constrain each other (outside of capacity-building or denying), and if so, how? That is, what are the channels of influence – e.g. power relationships, economic factors, ability to shape social conditions, and/or cultural ties – and how important are they? How do these co-evolve? Under which conditions have states found it in their favor to intentionally encourage and/or facilitate non-state actor acquisition of WMD? Have state actions in the past unintentionally created motivation for non-state actors to acquire WMD? What is the evidence base for state actions to have a deterrent affect (motivationally) on non-state actors to acquire of WMD? How do the relationships of primary state actors and secondary state actors to each other and to the cause of the non-state actor affect the choices made by non-state actors with respect to acquisition of WMD?

3 Some new questions present in the literature highlight the importance of second order effects, the larger set of actors that shape motivational factors, and the impact of unmediated communication channels (i.e. the internet, cell phones, electronic media distribution). Some of the important actors being considered are transnational entities such as NGOs, criminal networks and diaspora communities.[35] Research contributions from other disciplines provides a valuable knowledge base for examining strategic decisions to engage in mass violence and how these shape motivations to acquire (or not) WMD. Some new ideas that appeared from this broadened perspective suggest "barely-tapped" resources that can contribute meaningfully to an understanding of motivations for WMD acquisition by non-state actors:

   - consideration of more extensive data on mass violence and new methods of analysis of existing data;
   - transdisciplinary fields that integrate considerations of sacred values, culture, social psychology, and group dynamics;
   - the trade-off between non-violent actions and violent actions, and use of these in negotiation strategies, to compensate for asymmetric power relationships between actors engaged in struggle;
   - strategies for motivating mass support for movements, and creating perception of legitimacy and authority for challengers to state authorities;
   - de-escalation of spirals of mass violence;
   - the research on collective action and violence can be organized around questions of causal mechanisms that move individuals and groups from belief to action, from emotion to action, and combinations of the two.

4 Rigorous analyses of historic data on mass casualties may challenge conventional wisdom regarding which actors are most likely to employ WMD, as they show in the long view the relative ineffectiveness of this tactic to achieve objectives of non-state actors.

5 To understand the evolutionary dynamics of motivations, the interplay between competing motivations of actors and how trajectories toward or away from WMD proliferation are influenced by state actions, there needs to be more extensive research into second order effects with better exploitation of confluence of ideas from social science literature on religion and violence; radicalization; social movements; communications theory; anthropology; and history. Even with such understanding, however, the ability to incorporate into operational models of proliferation pathways with analysis tools currently available is questionable. Some means of analysis and new

databases have appeared that allow dynamic, non-linear analysis of these second order effects, but inclusion of complex models of motivation within these are at preliminary stages and are the exception, rather than the rule.

6 The potential for analysis of firsthand sources provided by extremists and potential terrorists themselves is only just beginning to be tapped for understanding the motivations and the debates that are incurred in developing intent to action. However, the methodologies for analysis need to incorporate controls for a number of factors such as intentional disinformation, framing of messages to adjust for intended audiences, sensationalism and credibility of sources.

A recurrent proposition which appears in the literature is that "if al-Qaeda could get their hands on WMD, they would use it." This proposition is often cited by the national security community as fact without consideration of caveats, context or critical examination of the data on which the statement is made. In fact, there was no evidence in the corpus of literature since 2001 to conclusively refute or confirm this statement. The study did reveal contradictory arguments that the evidence of history suggests political ju-jitsu is far more effective than mass violence for achieving goals of non-state actors. Strategic analysis of al-Qaeda documents themselves are ambiguous – revealing an ultimate emphasis on mass effect – such as winning the hearts and minds of the Muslim world, re-establishment of the Umma, and a risk-averse, conservative approach to choice of means and tactics. Mass destruction is called for as one of the means (out of many others – such as bringing down the US economy) called upon as acceptable, but not necessarily preferable, to achieve this.

At the same time, some of the complex dynamic interactions between exogenous and endogenous forces, tensions between competing motivations (which introduce vulnerabilities that can be exploited) and co-evolution of strategies to satisfy sometimes mutually exclusive goals are starting to be addressed. For example, there is increased attention being paid to understanding how the means and ends are inter-related and may constrain each other. This is true not only for al-Qaeda, but for all non-state actors who depend upon maintaining legitimacy in the eyes of a base of support in whose name they act.

There is an unresolved tension in the literature on which level is appropriate as a unit of analysis for these interactions. Those who study motivation and how it shapes behaviors differentiate between drivers at the individual, group and societal levels for when and how action is chosen. Within each level, the motivations are fairly well understood. However, the inter-relationship of drivers across these levels is not well understood. These inter-dependencies need to be accounted for in examining motivations for WMD acquisition. Cultural differences, especially between Western and Islamic contexts, also need to be accounted for.

There is little ongoing interaction evident in the literature amongst the communities of researchers, and between researchers and practitioners. However, this may be changing. For those in the social sciences who consider the question "How do non-state actors engage in struggles to achieve their goals, and how are decisions regarding violence and non-violence considered?" there are some who look at the possible use of WMD. This in turn is beginning to expand (albeit by marginal increments) the social science community actively engaged in the question of WMD proliferation.

Implications

There are three primary implications suggested by this research. The first is that improvements are needed in the research methodologies for valid forecasts of WMD acquisition by non-state actors. The second is that more interdisciplinary research questions should drive studies of motivations. The third is that there is a need for an integrating framework for analyzing WMD decisions which accommodates diverse disciplinary knowledge to be integrated without losing rich details of behavior mechanisms from each area.

Research methodologies

Deficiencies in research methodologies represented in the literature present cause for concern in the internal and external validity of analyses, judgments and predictions. These deficiencies include a continued, unchallenged prevalence of "conventional wisdom" in the literature for explaining WMD motivations of non-state actors, the isolation of enclaves of experts among academia and practitioners of WMD threat analysis, and the paucity of structured, evidentiary-based reasoning. In assessing the validity of scientific forecasts by experts, Dr. J Scott Armstrong of the Wharton Business School, has shown that the accuracy of unstructured analyses and/or judgments by experts are no more accurate than novices (in some cases, high-school students) for predicting outcomes and behaviors (Armstrong 2001). In one study based on 1,736 predictions about consumer behaviors, none of the subject groups performed better than chance. Furthermore, the experts performed much worse than they or their peers had predicted they would, implying an inflated self-evaluation of the merit of their forecasts (Armstrong 1991). Possible explanations for poor forecasts are the lack of structure in the experts' analysis and the nature of the scientific research itself. Armstrong's subsequent studies have shown that when experts are aided by structured analytic methods appropriate to the problem, forecasts improve.[36] However, consistent with principles of validity in conducting social science research, he also argues that these forecasts may continue to be inaccurate unless based on research that yields generalizable findings, that the generalizations do not yield unambiguous predictions, that the findings are effectively written, that the researchers themselves have understood the full scope of all relevant science and (1) are willing to believe the findings and (2) know how to effectively use and communicate the knowledge. The current research base on WMD suffers in these areas.

The efforts underway to validate and make large databases public (such as at the University of Maryland) and the growing ability to search across multiple databases are positive developments toward a broader empirical basis for analyzing motivations. These developments could become much more powerful if supplemented by more appropriately structured analysis processes. However, these need to be carefully chosen based on the complexity of the analysis as measured through structural and dynamical considerations, the epistemology of the analysis – what knowledge is to be generated and how it is to be used, and the information density required, depending on the timescale of the analysis relative to action required and data available (Hayden 2007).

In cases where relationships and interactions between actors are a primary concern, and data is plentiful, social network analysis provides a meaningful structural aid to analyzing motivations and considering how they are influenced and shaped. Alternately, if the primary question involves learning and adaptation mechanisms experienced over time, neural or Bayesian networks are more appropriate. Where data is lacking, Armstrong has shown that structured reasoning using analogues can lead to highly credible and accurate forecasts.[37] Another structured analysis method useful for hypothesis testing where data is unavailable is agent-based modeling, which can provide simulated data conditional upon the hypothesized conditions. Agent-based models are also useful when the underlying phenomena of interest is emergent, with complex behaviors often resulting from simple interactions. The models can be used to isolate those interactions which matter most, using stochastic processes to generate probabilistic estimates.

Interdisciplinary research questions

Analysis of the literature showed additional questions that might be addressed by future research. Some of these are:

1. What are the relationships between deterrence methods and motivations? How do they shape each other in a dynamic sense, and what types of influencing feedback loops are created by different policy strategies?
2. What evidence exists for understanding the relationship between the base of sympathetic support for groups and the moderating or amplifying effect on motivations to acquire and/or use WMD?
3. What are the different models for understanding how external motivational drivers (i.e. resource mobilization, political opportunity structures, repression) and internal drivers (i.e. inter-group competition, leadership challenges, alienation) interact to shape decisions regarding acquisition and/or use of WMD? How are these contextually and culturally dependent, and socially and psychologically specific?
4. Under what conditions have non-violent repertoires of contention been employed by non-state actors as alternatives to the same motivational factors that exist for WMD acquisition and/or use?
5. Under what conditions have non-state actors desisted in the pursuit of WMD to achieve goals?

6 What are additional research perspectives that can be brought to bear on the question of motivation of acquisition and/or use of WMD from experts in transnationalism, globalization and world systems?

Integrative frameworks

When incorporating motivation into a systems' level model of non-state WMD proliferation, frameworks that integrate mechanisms across disciplines should be provided for exploring alternatives to WMD available to non-state actors that include not only conventional means of violence, but also non-violent actions that have historically been brought to bear to address socio-political grievances and power struggles against oppressive forces.

By doing so, resultant models can better accommodate not only behaviors that may directly deter terrorists and extremists groups, but also allow the consideration of second order effects that include the populations in whose name the non-state actors act, who hold these grievances. This understanding is critical for present and future US national security policy initiatives that emphasize strategic communication. The disenchantment with US policies among Muslim communities and the increasing numbers of populations worldwide – including allies with large diaspora communities – make it imperative that these issues continue to be explored.

An example of such an integrative framework might be one that is developed around internal and external drivers of cohesion and power structures, respectively. As pointed out in the literature on non-violent campaigns, internal debates about violence and non-violence and the degrees of acceptable violence have been a key factor in the disintegration of opposition groups. Further work to explore such integrative frameworks would be a valuable contribution toward bridging between academic studies and policy-making, *when grounded in underlying theory and data*. Too often frameworks are posed that are narrow in theoretical scope, or lacking theory altogether.

Summary

Understanding motivations of non-state actors for acquisition and/or use of WMD is vital for effective national security and foreign policies. Researchers and policy-makers alike share the responsibility of ensuring that the knowledge base which supports such understanding meets basic criteria, first proposed by Alexander George for decision-making in this arena (George 1980):

1 Ensure that sufficient information is obtained and analyzed adequately to provide a valid diagnosis of the problem.
2 Consider all the major values and interests affected by policy issue at hand.
3 Assure a search for a relatively wide range of options, with consideration of costs, risks, and benefits of each.

If these conditions are met, one can imagine that the landscape of knowledge presented by the research underlying the decision basis would resemble a rich and diverse mixture of woodlands and fields, with perhaps a few relatively high peaks where particularly relevant material is dense. Instead, today's research into WMD motivations of non-state actors generates a landscape that is more reminiscent of Monument Valley, in the desert areas of Utah. Here, isolated buttes stand high in stark contrast to the flat and empty desert floor separating the monolithic structures. Over time, one hopes that the landscape will flourish.

Notes

1 According to Bruce Hoffman, no terrorist attack prior to 9/11 had ever killed more than 500 people. In the twentieth century only 14 events have killed more than 100 people (Hoffman 2007).
2 On May 11, 2008, the Russian news service, RIA Novosti, reported that Russia's antiterrorism committee had said it had evidence that terrorists were trying to gain access to weapons of mass destruction and to technology needed to produce them.
3 For example, Raphael Perl has summarized to the US Congress the public media accounts (of varying credibility) suggesting that Usama bin Laden has joined the WMD procurement game. A London *Daily Telegraph* dispatch (December 14, 2001) cites "long discussions" between bin Laden and Pakistani nuclear scientists concerning nuclear, chemical and biological weapons. The *Hindustan Times* (November 14, 2001) claims that a bin Laden emissary tried to buy radioactive waste from an atomic power plant in Bulgaria and cites the September 1998 arrest in Germany of an alleged bin Laden associate on charges of trying to buy reactor fuel (see also *London Times*, 14 October 2001). A US federal indictment handed down in 1998 charges that bin Laden operatives sought enriched uranium on various occasions. Other accounts credit al-Qaeda with attempting to purchase portable nuclear weapons or "suitcase bombs" through contacts in Chechnya and Kazakhstan. Furthermore, US government sources reported discovery of a partly constructed laboratory in Afghanistan in March 2002, in which al-Qaeda may have planned to develop biological agents, including anthrax. In April 2002, a captured al-Qaeda leader, Abu Zubaydah, told American interrogators that the organization had been working aggressively to build a dirty bomb, in which conventional explosives packaged with radioactive material are detonated to spread contamination and sow panic. BBC reports (January 30, 2003) cite the discovery by intelligence officials of documents indicating that al-Qaeda had built a dirty bomb near Herat in Western Afghanistan. In January 2003, British authorities reportedly disrupted a plot to use the poison ricin against personnel in England (Perl 2005).
4 In 2001 Jonathan Tucker outlined the ease of proliferation of chemical and biological materials and technologies to state and sub-state actors in testimony before the US Senate (Tucker 2001). More recently, the US State Department Office of Counterterrorism has noted that: (1) the diffusion of scientific and technical information regarding the assembly of nuclear weapons, some of which is now available on the internet, has increased the risk that a terrorist organization in possession of sufficient fissile material could develop its own crude nuclear weapon; (2) the scientific capabilities for biological weapons are not beyond the expertise of motivated biologists with university-level training and that the materials are widely available; and (3) the growth and sophistication of the worldwide chemical industry makes the task of preventing and protecting against chemical weapons difficult, and that terrorists could use commercial industrial toxins, pesticides and other commonly available chemical

agents as low-cost alternatives to militarized weapons and delivery systems, though with limited effects (US State Department 2008).

5 The exploitation of media is critically important to both the non-state actors who might use WMD, and to government officials promoting political agendas. In the case of the latter, ethical use of the media can accurately inform publics of threats and how to respond. Unethical, unmediated use can overinflate threat and propagate terror. This is an increasing consideration with the global reach of diverse news venues with different standards and perspectives (Nacos 2007)

6 Thomas Homer-Dixon notes,

> Modern societies face a cruel paradox: fast-paced technological and economic innovations may deliver unrivalled prosperity, but they also render rich nations vulnerable to crippling, unanticipated attacks. By relying on intricate networks and concentrating vital assets in small geographic clusters, advanced Western nations only amplify the destructive power of terrorists – and the psychological and financial damage they can inflict.
>
> (Homer-Dixon 2002)

7 A notable example is the work of Alexander George at Stanford University in his studies of stress and political decision-making (George 1980). In these studies, he drew both on persons with practical experience in foreign-policy making and on academic specialists in behavioral sciences – psychologists, political scientists, sociologists, economists, organizational theorists and decision theorists – who utilized many different kinds of data and employed a variety of research methods.

8 Research literature was restricted to professional journals and other sources that ensure a peer-review process. This may be a serious impediment for incorporating non-Western views, as the channels for conducting and publishing research differ in non-democratic societies, and often preclude the type of peer-review processes that are standard in the West.

9 In referring to the rational actor decision-making model, the classical paradigm given by Graham Allison is presumed, wherein the rational actor is a unified, individual unit, however parsimoniously or prescriptively described (ranging from some notional unit such as a state or non-state actor, to a generic unit such as a democratic government or a terrorist group, to a specified unit such as the United States or al-Qaeda, to a specific individual such as George W. Bush or Usama bin Laden. The actor's actions are explained or predicted in terms of the objective conditions it faces, combined with four variables in the concept of rational action alone: objectives, options, consequences and choice. Choice is made by maximizing value that can be prescribed in a coherent utility function of the four variables. Allison points out that a rigorous analysis using the rational actor paradigm must insist on "rules of evidence for making assertions about objectives, options, and consequences that permit distinction among various accounts" to avoid the trivial case wherein any pattern of activity can be explained by imaginatively constructed objective functions for actors" (Allison 1999, p. 26).

10 A number of other definitions of WMD can be found at www.answers.com/topic/weapons-of-mass-destructionc/opyright.

11 See, http://nti.org/f_wmd411/f1a1.html for a list of documents that include Presidential directives, US State Department policy, DoD doctrine and Congressional legislation.

12 Statement of Major General Robert P. Bongiovi, USAF, Acting Director, Defense Threat Reduction Agency, Before the Emerging Threats and Capabilities Subcommittee, Committee on Armed Services United States Senate, July 12, 2001.

13 However, as will be pointed out in the results of the study, there are motivational factors for making the choices between technologies that have to do with perceptions of risk and propensity for risk-taking behaviors.

14 Theories to explain motivation are derived primarily from behavioral, social and cognitive psychology literature. The earliest and most basic needs theories, pioneered by

Maslow, explain motivation according to a hierarchy of needs: physiological, security/safety, belongingness and love, esteem, self-actualization and transcendence (Maslow 1943). Expansions on Maslow's work bring in more complex ideas of the development of needs in relation to the environment and others, such as deprivation and attribution theories. Process theories explain how individuals select particular behaviors and how individuals determine if these behaviors meet their needs. The two primary drivers in these theories are expectancy and perceptions of equity. Theories of reinforcement, pioneered by Skinner, are based not on need but on the relationship between behavior and its consequences. According to this theory, reinforcements can be positive, negative, punishment or extinction.

15 Access to special collections and subject matter expertise through institutional resources at Sandia National Laboratories provided guidance for expanded search terms in the public domain.

16 In some cases, allowance was made for citations that, while lacking substantial data, posed new lines of inquiry of interesting, but as yet untested, hypotheses.

17 Two software packages developed at Sandia National Laboratories (TaMALE and STANLY) and one commercial text extraction middleware product available for federal systems (InXIGHT) were used.

18 Four levels of relevance to WMD were considered. The most direct were those that used the term WMD and were intentionally focused on explanations for its acquisition and/or use. The second level was literature that examined motivations for use of technologies that are typically associated with WMD (e.g. biological warfare agents, nuclear materials) to cause mass destruction; the third level was literature that dealt with motivations to use or threaten to use mass violence and/or destruction in general against non-combatant civilians to achieve a goal; the fourth was literature that dealt with decision-making between violence and non-violence by non-state extremists to engage in social and political struggle.

19 See, for example, Stern (1999) or Roberts (1999).

20 An exception is the potentially rich source of data becoming more publicly accessible on nuclear smuggling, such as the Database on Nuclear Smuggling, Theft, and Orphan Radiation Sources (DSTO) operated by the University of Salzburg in Austria.

21 Ongoing studies with the GTD can be found at www.start.umd.edu/data/gtd.

22 Online at www.johnstonearchive.net/terrorism/wrjp394.html.

23 Bibliographic sites included in data-mining research for this chapter are maintained by the following organizations: the American Political Science Association, the US Joint Forces Staff College, the US Naval Postgraduate School, the Maxwell Air Force Base, the US Military Academy, the US National Defense University Center for the Study of Weapons of Mass Destruction, the Center for Nonproliferation Studies, Jane's Intelligence Review, the US Department of State, the US Department of Homeland Security, the US Central Intelligence Agency, Harvard University, Library of the US Congress Federal Research Division, the Nuclear Threat Initiative, the North Atlantic Treaty Organization, the Memorial Institute for the Prevention of Terrorism and the Commonwealth Institute.

24 See for example, Stohl (2005) on expected utility theory, or Lichbach (2003) on the bargaining theory of war and the rebel's dilemma as decision-making models, compared to work by Atran (2006), Atran *et al.* (2007) and Sosis and Alcorta (2007) on the motivational role of sacred values.

25 See, for example Ginges *et al.* (2007).

26 This has been a high research-growth area in the past six years, and a complete bibliography will not be presented in this chapter. However, for a sense of the growth in the field, in addition to the citations already noted, one might examine the work of Dishman (2001) or Auerswald (2006).

27 Statistically significant data has been hard to come by for such global studies in the past. With the public availability of the GTD, there may be more opportunity

to study the interaction of these long-term, global trends and motivation factors for WMD.

28 See for example, the analysis of complex new geopolitical relations between Saudi Arabia and various Islamic groups by Lacroix (2004) and the studies on trends in SE Asia reported by Tan (2003).

29 The full set of reports and original source materials can be found at http://ctc.usma.edu/publications/publications.asp.

30 One such notable scholar is Bruce Hoffman.

31 Personal communication, Professor C.M. McCauley, Bryn Mawr College, 2004.

32 The clique model of radicalization among the 9/11 perpetrators was put forth by Marc Sageman in numerous professional presentations in the months before publication of his book, *Understanding Terror Networks* (2004) and is carried forward in his most recent publication, *Leaderless Jihad* (2007).

33 See, for example, McAdam *et al.* (2001).

34 Parachini (2003) notes that complex factors shape a group's propensity to acquire and use unconventional weapons, and that while religion is an important one, it is not the only one. The interplay between motivation, opportunity and technical capability is too often overlooked. As an example, he cites the efforts of the IRA and FARC to collaborate on chemical weapons. These efforts were eventually abandoned due to constraints, not necessarily lack of motivation. Such interplay is hard to observe and assess accurately.

35 Much of the new literature on minority communities addresses Muslim diaspora. However, there is some new research drawing on historical case studies of interactions between state responses to terrorism and minority community reactions, such as Ramraj (2006).

36 Armstrong maintains an up-to-date methodology for selecting structured methods of analysis at his website, www.forecastingprinciples.com/selection_tree.html.

37 Armstrong presents a decision-making tree for choosing methods based on data availability and degrees of freedom at www.forecastingprinciples.com/methodologytree.html.

References

Ackerman, Gary (2005) "WMD Terrorism Research: Whereto From Here?" in Andrew Blum, Victor Asal and Jonathan Wilkenfeld (eds.) "Nonstate Actors, Terrorism and Weapons of Mass Destruction," *International Studies Review*, 7: 140.

Allison, Graham (1999) *Essence of Decision*, New York: Addison-Wesley Educational Publishers.

Allison, Graham and Philip Zelikow (1999) *Essence of Decision: Explaining the Cuban Missile Crisis*, New York: Addison Wesley, Longman Inc.

Armstrong, J. Scott (1991) "Prediction of Consumer Behavior by Experts and Novices," *Journal of Consumer Research*, 18: 251–6.

Armstrong, J. Scott (2001) *Principles of Forecasting*, New York: Springer.

Asal, Victor and Andrew Blum (2005) "Holy Terror and Mass Killings: Reexamining the Motivations and Methods of Mass Casualty Attacks," in Andrew Blum, Victor Asal and Jonathan Wilkenfeld (eds.) "Nonstate Actors, Terrorism and Weapons of Mass Destruction," *International Studies Review*, 7: 154.

Atran, Scott (2006) "Failure of Imagination," *Studies in Conflict and Terrorism*, 29, 3: 263–83.

Atran, Scott, Jeremy Ginges, Douglas Medin and Khalil Shikaki (2007) "Sacred Bounds on Rational Resolution of Violent Political Conflict," National Academy of Sciences, available at: www.pnas.org/content/104/18/7357/abstract.

Auerswald, David (2006) "Deterring NonState WMD Attacks," *Political Science Quarterly*, 121, 4: 543–68.

Bell, Joyce (2003) "The Cultural and Structural Determinants of Social Movement Factionalism and Radicalization," ASA Annual Meeting 2003, Atlanta, Georgia.

Center for Nonproliferation Studies (2002) "Literature Review of Existing Terrorism Behavior Modeling: Final Report to the Defense Threat Reduction Agency," Defense Threat Reduction Agency, Washington DC. Available at: http://cns.miis.edu/pubs/reports/terror_lit.pdf.

Daft, Richard L. (1997) *Management*, 4th edn, Orlando, FL: Harcourt Brace.

Dishman, Chris (2001) "Terrorism, Crime, and Transformation," US Commission on National Security.

Dolnik, Adam (2003) "All God's Poisons: Re-Evaluating the Threat of Religious Terrorism with Respect to Non-Conventional Weapons," Monterey Institute International Studies Report Prepared for the Defense Threat Reduction Agency. Published in *Terrorism and Counterterrorism* (McGraw-Hill) in 2005.

Drennan, Shane (2006) "Fourth Generation Warfare and The International Jihad," *Jane's Intelligence Review*.

Ferraro, Mario (2005) "Radicalization as a Reaction to Failure: An Economic Model of Islamic Extremism," *Public Choice*, 122, 1–2: 199–220.

George, Alexander (1980) *Presidential Decision-Making in Foreign Policy: The Effective Use of Information and Advice*, Boulder, CO: Westview Press.

Ginges, Jeremy, Scott Atran, Douglas Medin and Khalil Shikaki (2007) "Sacred Bounds on Rational Resolution of Violent Political Conflict," *Proceedings of the National Academy of Sciences*, 104: 7357–60.

Goodwin, Jeff (2006) "A Theory of Categorical Terrorism," *Social Forces*, 84, 4: 2027–46.

Hayden, Nancy K. (2007) "The Complexity of Terrorism: Social and Behavioral Understanding Trends for the Future," in Magnus Ranstorp (ed.), *Mapping Terrorism Research: State of the Art, Gaps, and Future Directions*, London and New York: Routledge, pp. 292–315.

Hoffman, Bruce (1997) "Terrorism and WMD: Some Preliminary Hypotheses," *Nonproliferation Review*, 104, 43: 45.

Hoffman, Bruce (2007) "CBRN Terrorism Post-9/11," in Russell D. Howard and James J.F. Forest (eds.), *Terrorism and Weapons of Mass Destruction*, New York: McGraw-Hill.

Homer-Dixon, Thomas (2002) "The Rise of Complex Terrorism," *Foreign Policy*, 128: 56–62.

Lacroix, Stephane (2004) "Between Islamists and Liberals: Saudi Arabia's New 'Islamo-Liberal' Reformists," *Middle East Journal*, 58, 3: 345–65.

LaFree, Gary, Laura Dugan and Derrick Francke (2005) "The Interplay Between Terrorism, NonState Actors, and Weapons of Mass Destruction: An Exploration of the Pinkerton Database," in Andrew Blum, Victor Asal and Jonathan Wilkenfeld (eds.) "Nonstate Actors, Terrorism and Weapons of Mass Destruction," *International Studies Review*, 7: 156.

Leven, Mark (2002) "The Changing Face of Mass Murder: Massacre, Genocide, and Post-genocide," *International Social Science Journal*, 54, 4, 174: 443–52.

Lichbach, Mark Irving (2003) *Is Rational Choice Theory all of Social Science?*, University of Michigan Press.

McAdam, Doug, Sidney Tarrow and Charles Tilly (2001) *Dynamics of Contention*, New York: Cambridge University Press.

McCants, William and Jarrett M. Brachman (2006) "The Militant Ideology Atlas," *Harmony Report*, Combating Terrorism Center, US Military Academy. Available at: http://ctc.usma.edu/publications/publications.asp.

Mallon, Will (2008) "WMD: Where Did the Phrase Come From?" George Mason University History News Network. Available at: http://hnn.us/articles/1522.html (accessed April 2, 2008).

Maslow, A.H. (1943) "A Theory of Human Motivation," *Psychological Review*, 50: 370–96.

Nacos, Bridgette (2007) *Mass-Mediated Terrorism: The Central Role of the Media in Terrorism and Counterterrorism*, Lanham, MD: Rowman & Littlefield Publishers, Inc.

Parachini, John V. (2001) "Comparing Motives and Outcomes of Mass Casualty Terrorism Involving Conventional and Unconventional Weapons," *Studies in Conflict and Terrorism*, 24: 389–406.

Parachini, John V. (2003) "Putting WMD Terrorism into Perspective," *Washington Quarterly*, 26, 4: 37–50.

Perl, Raphael (2005) "Terrorism and National Security: Issues and Trends," CRS Issue Brief for Congress, Order Code IB10119.

Post, Jerrold (2000) "Psychological and Motivational Factors in Terrorist Decision Making: Implications for CBW Terrorism," in J. Tucker (ed.), *Toxic Terror*, Cambridge, MA: MIT Press.

Ramraj, Victor Vridar (2006) "Counter-Terrorism Policy and Minority Alienation: Some Lessons from Northern Ireland," *Singapore Journal of Legal Studies*, 385–404.

Rittel, Horst and Melvin Webber (1973) "Dilemmas in a General Theory of Planning," in *Policy Sciences*, volume 4, Amsterdam: Elsevier Scientific Publishing, pp. 155–9.

Roberts, Brad (ed.) (1999) "New Terrorism: Does it Exist? How Real Are the Risks of Mass Casualty Attacks?" CBACI Conference Report.

Roberts, Brad (2002) "Motivation for Terrorists to Use Weapons of Mass Destruction," *Confronting Terrorism*, a workshop held at Los Alamos National Laboratory March 25–29, 2002. Proceedings edited by Rajan Gupta and Mario R. Perez.

Sageman, Mark (2004) *Understanding Terror Networks*, Philadelphia, PA: University of Pennsylvania Press.

Sageman, Mark (2007) *Leaderless Jihad: Terror Networks in the Twenty-First Century*, Philadelphia, PA: University of Pennsylvania Press.

Sheffer, Gabriel (2005) "Diasporas, Terrorism, and WMD," *International Studies Review*, 7, 1: 160–2.

Shelly, Louise (2007) "Growing Together: Ideological and Operational Linkages Between Terrorist and Criminal Networks," The Fund for Peace Expert Series, Washington, DC.

Sosis, Richard and Candace Alcorta (eds.) (2007) "Adaptive Militants and Martyrs? What Evolutionary Theories of Religion Tell Us About Terrorists," in Rafe Sagarin and Terrence Taylor (eds.) *Darwinian Security: Perspectives from Ecology and Evolution*, Berkeley, CA: University of California Press.

Steinbruner, John (2005) "Terrorism: Practical Distinctions and Research Priorities," *International Studies Review*, 7, 1: 137–40.

Stern, Jessica, (1999) "Prospect of Domestic Bioterrorism," *Infectious Diseases*, 5, 4: 517.

Stohl, Michael (2005) "Expected Utility and State Terrorism," in Bjo Torero (ed.) *Root Causes of Terrorism: Myths, Realities, and Way Forward*, New York: Routledge, pp. 189–97.

Tan, Andrew (2003) *The New Terrorism: Anatomy, Trends, and Counter-Strategies*, Singapore : Eastern Universities Press.

Tucker, Jonathan B. (2001) "The Proliferation of Chemical and Biological Weapons Materials and Technologies to State and Sub-State Actors," Testimony before the Subcommittee on International Security, Proliferation, and Federal Services of the US Senate Committee on Governmental Affairs, Washington, DC.

US State Department (2008) "The Global Challenge of WMD Terrorism," in *Country Reports on Terrorism*, chapter 4. Available at: www.state.gov/s/ct/rls/crt/2007/103712.htm (accessed May 11, 2008).

10 Conclusion

Magnus Ranstorp and Magnus Normark

At the end of May 2008, the Al-Ikhlas forum posted a 39-minute video entitled *Nuclear Jihad, the Ultimate Terror*, in which it implied an imminent nuclear strike on the West. The video, audio and text contained a veritable collage of interwoven emotional appeals to defend against the West's aggression against Muslims; Abu-Musab al-Suri's appeals for Muslims to rise up and leave the United Kingdom, as well as a multi-forum dialogue about the need to wage jihad using CBRN agents and weapons. While this served as a stark reminder of the strategic threat posed by CBRN terrorism and the necessity for counter-proliferation efforts, there is still immense uncertainty as to the nature of threat convergence between CBRN and terrorists groups, transforming theory into reality.

As the US journalist I.F. Stone used to say "if you expect an answer to your question during your lifetime, you were not asking a big enough question."[1] This volume has tried to ask a "big question" as to what multiplicity of factors are at work both in terms of terrorist groups behavior, incentives and disincentives and the technical dimensions providing barriers and possibly opportunities in an age of globalization and immense technological change. If we are to believe Chris Donnelly, former adviser to four NATO secretary generals in succession, Western societies are now at war and in revolution, though we have, and operate with, a peacetime mentality. According to Donnelly, the definition of this war and revolution is the *speed of change* both in terms of technology and society.[2] Predicting the speed of change and the multilayered conflation of the global and the local through the effects of globalization is truly a complex and even "wicked problem" as originally defined by Horst Rittel and Melvin Webber. Framing the problem correctly and illuminating multiple angles is essential to understanding and resolving "wicked problems."[3]

In this volume we have tried to frame a series of questions that disaggregates the complexity and dimensions of the problems of the convergence when terrorist groups will actually decide to pursue CRBN capability offensively. This disaggregating exercise occurred along two axis – the terrorist group dimensions, on the one hand, which combines motivation, group dynamics, decision-making and ultimately tactical as well as strategic innovation, and on the other hand, the technical dimensions and dynamics according to each agent, as well as varying obstacles and opportunities in the procurement, production and dissemination processes. An integral part of the convergence process connecting these two dimensions rests on applying the appropriate conceptual level in thinking about the problem. The basic premise for this study was that the existing literature had reached an intellectual impasse, in the words of Gary Ackerman – who in many ways was the intellectual conductor for this project – and there was a need to think in new ways about and around the problem. Hopefully this volume provides new avenues to understand the complexity of the interlocking factors affecting threat convergence. This is a hugely complex issue that has multiple analytical layers and is affected by constantly changing dynamics within terrorist groups and the enabling environment. Modestly, the principal aim of this study has been to provide analytical pathways that can hopefully provide or stimulate new ways of thinking about threat convergence.

What then are the central features?

Terrorists attitudes toward CBRN – incentives (and disincentives)

Understanding social and behavioral aspects of terrorist groups and their decision-making processes is notoriously difficult – more of an art than a science, despite advances in modeling complex social networks. In efforts to capture terrorist behavior, several research studies have tried to frame the nature of the threat environment in general, specifically identifying key factors behind decisions to escalate the violence levels according to motivation, ideology or crisis situations, forcing groups toward behavioral anomalies or leaps in the level and intensity of violence. The excellent volumes by Jonathan Tucker and John Parachini[4] provided new and important insights through a series of known historical case studies into what factors influenced and provided incentives for groups to pursue CBRN pathways. Importantly, these individual case studies were highly granulated and this type of fact-finding and analysis provided useful means to dispel inaccuracies that often circulated in the academic literature and media about CBRN acquisition without being challenged or corrected. In addition, other studies have focused on escalation dynamics toward levels of mass destruction within terrorism, as well as assessment of the various potential technical pathways toward such attacks. As Gary Ackerman has accurately pointed out, there is a division between studies and assessments of the threat from primarily two different disciplines. One highlights the technical features of pursuing a CBRN terrorism capacity and the potential consequences of such events. The other perspective focuses on the behavior of sub-state actors, their ideologies, ambitions and past modus operandi as indicators for future CBRN behavior.

Most social science-based studies have to a large extent explored the possible *incentives* behind terrorists' intent and ambitions to acquire a CBRN capability in an effort to predict the profile of the actors that might be most prone toward being committed to a CBRN pathway for future terrorist acts. These studies are indeed important for several reasons. Obviously it is a necessary pathway to explore as law enforcement and intelligence communities as well as policy-makers try to get a handle on a low risk but high impact phenomenon that has yet to occur. The study of incentives also serves as an important avenue to identify early warning indicators of future attempts to deploy terror attacks with CBRN components. However, there are related dimensions which could contribute to new insights of the threat and the work in countering the risks of future CBRN terrorism attacks that haven't been explored in any great detail. Just as it is important to ask what incentives exist for the pursuit of CBRN pathways, it is equally illuminating to ask what *disincentives* exist that primarily influence why terrorists stick conservatively to traditional means and methods. Clearly, studying these aspects from a disincentives perspective would complement and further our understanding of the CBRN threat significantly, both in the short- as well as long-term perspective. In fact, we would argue that focusing on disincentives could be more revealing than on incentives, considering the fact that terrorist groups are becoming less hierarchical, more networked and innovative, which would suggest propensity toward technical and methodological innovation. The fact that this diffusion and decentralization of structure, combined with new opportunities accorded by the forces of globalization, have not resulted in a significant transformation of the terrorists' modus operandi and expanded toolbox for violence is illustrative of the necessity to focus on *disincentives.*

Future research would gain considerably by focusing on mapping the dynamics of disincentives: What is keeping the contemporary terrorists from entering a CBRN pathway in their efforts to reach their objectives? What can influence the current disincentives in the future, making terrorists more prone to choose CBRN as their violent means? Which aspects of possible disincentives should be re-enforced or enhanced as a part of preventive strategies against CBRN terrorism?

Similarly more research is necessarily prudent on incentives as well. What attracts different kinds of terrorist cells to CBRN? The prospect of large-scale violence? The "unconventional" nature of weapons and the fear it creates in a victim society? A weapon system better suited to political goals than "conventional" weapons? Does al-Qaeda central really feel the need to raise the level of violence with every centrally planned attack? How does it weigh the psychological value of using CBRN versus the possibly greater physical impact of conventional weapons? Has the general debate regarding threat perception of CBRN terrorism from media, governments, institutes and non-governmental organizations during the early twentieth century influenced the terrorists' attitude toward CBRN as a means for enhancing the effect of their actions? How do we make societies more resilient to CBRN attacks? In other words, how do we communicate with the public and give them appropriate tools for dealing with an attack? Do terrorist groups that aim to produce political effects with their attacks take such resiliency into account when planning an attack? In other words, if a society is more resilient to a CBRN attack, will terrorist organizations find such attacks less valuable?

Answering these critical questions is a complex undertaking indeed, but a central issue is to better understand how terrorist groups *innovate*. Apart from Charles Drake's classic study on terrorist targeting, there are few studies focusing on the organizational dynamics that drive innovation and adaptation within and between groups. A recent valuable contribution on terrorist learning process has been made by Horacio Trujillo and Brian Jackson.[5] This study differentiates between incremental and transformational learning across different determinants of organizational learning: structure; culture; knowledge resources; and environment.[6] In terms of structure, the study points to research that suggests that "more hierarchical organizations frequently learn less effectively, due in part to the loss of information as it is transmitted through and screened by the different organizational levels."[7] Similarly, so-called hybrid structures – decentralized network structures – often need to spend more time on management procedures to maintain security. On the cultural level, Trujillo and Jackson argue that there is "two interrelated and mutually reinforcing cultural traits that are conducive to organizational learning – *organizational interest in learning and organizational tolerance for risk-taking.*"[8] On the other hand, there are constraints on learning "such as *role constraint, situational and fragmented learning* and *opportunistic learning.*"[9] And finally, the study suggests that *absorptive capacity* plays a key role, especially in the technology acquisition areas while "environmental uncertainty can spur organizational experimentation" and "*crises in the environment*" can lead to dramatic learning.[10] This approach provides a useful method to further study the modalities of terrorist behavior and innovation across different contexts.

An interrelated issue is better understanding of terrorist decision-making. A detailed understanding of the decision-making process would help identify the impediments a terrorist group would face for certain types of attacks. For example, for nuclear attacks, in assessing this decision process, one would find that the group would need a certain threshold level of money and technical skills as well as insider information and assistance about facilities containing nuclear weapons or nuclear materials. These impediments that the group would need to surmount would then point to certain indicators, for instance, the need to look for groups with certain skill levels and access to adequate amounts of money, etc.

Finally, there is the issue of *dynamic deterrence* to try to influence and shape terrorist decision-making. What are the lessons from the Cold War in shaping deterrence and can we adapt dynamic deterrence on both a sub-state and state level and construct accordingly? Is it possible to shape deterrence toward non-state actors effectively?[11] A RAND Study into *Deterrence and Influence in Counterterrorism* (2002) suggests that deterrence may have some applicability in the CBRN realm, but it is quite difficult as these contemporary terrorist groups are complex adaptive systems.[12] As Wyn Bowen has summarized this difficulty:

> the real challenge in determining whether nonstate actors like al Qaeda are susceptible to deterrence logic involves penetrating their black boxes. This means understanding the frame of reference of actors, how it is evoked, options considered in decisionmaking, and the lens through which they will perceive deterrent messages. Specifically, there must be emphasis on evaluating how specific groups or individuals calculate costs and benefits: Are

they risk prone or risk averse? Do they think in terms of minimizing losses or maximizing gains? To what extent are they motivated by survival, security, recognition, wealth, power, or success?[13]

The technical dimensions

Any social and behavior model assessing terrorist groups' motivation and intent toward a CBRN pathway must be weighted in relation to the specific competencies and technical assets that are within reach for them in order to assess present or future CBRN terrorism threats and risks. In an age of globalization, this identification task is complex given the polymorphous nature of terrorist networks that are motivated by extreme religious ideologies which legitimize violence on an ever-increasing scale, and possess no moral compunction of using CBRN in their attacks on society. This threat exists not only in the physical realm, but also on the internet where new alignments and constellations are formed and dissolved. This makes this threat very hard to detect and counteract as the threat environment is compressed from the global to the local in time and space.

The increasing difficulties in monitoring and controlling the transfer and storage of potentially dangerous CBRN materials and agents, not only on the international market but also within our own societies, makes the task of assessing future terrorist threats and risks increasingly difficult and complex. The technical perspective in the work of gaining further insights into potential terrorism threats and risks is not an uncontroversial dimension. There are numerous studies conducted by technicians that elevate the forecast of future threat environments and opportunities facilitating non-state actors' acquisition, development and deployment of biological or chemical agents through advanced and novel techniques.[14] Efforts to assess the technological development curve without taking into account the thresholds and disincentives from a social and behavioral standpoint are often poor and misleading guides to future potential terrorism threats. Does motivation really equal intent? We need to gain a better understanding of how the terrorists themselves perceive CBRN pathways to violence, not as we perceive the technical revolutions and their impact on opportunities to commit mass-impact attacks (terrorist cost–benefit calculations). One possible fruitful area is looking into terrorist *financing*, since this is a key to all CBRN threats; export controls and infrastructure requirements are also a key dimension to most CBRN threats. In short, the technical dimension is vital in assessing capabilities and potential pathways for CBRN terrorism as well as consequences and vulnerabilities in our societies to such events. However, this analysis is not without pitfalls.

Separating myths from facts

If rhetoric and intent by terrorist groups to deploy CBRN means are regarded as vital indicators of future terrorism events, then we are in serious trouble indeed. It is common knowledge that assessing the terrorist threat requires substantial granulated insight in both terrorists' intent and capabilities to conduct violent acts. It is, however, appalling how many references there are to terrorists' rhetoric and stated interests for deploying CBRN weapons without evaluation or benchmarking these declarations against capability aspects in the environment in which they may operate. Furthermore, the few attempts by non-state actors to acquire and deploy CBRN weapons have to a very large extent been unsuccessful and the issue of CBRN terrorism in the open debate is flawed with unsubstantiated allegations of highly unlikely events. One example of ambiguous references of WMD activities by non-state actors is the CNS chart on al-Qaeda's WMD activities.[15] An introduction to the list does provide a statement that the chart is based on a wide array of sources whose reliability is varied and with information that is uncorroborated. Nonetheless, the publication and reference to an extensive list of alleged WMD incidents, which is laden with reports that are highly unlikely and contain duplicates and cross references to the same incident, provides for the uninitiated a grossly overstated account of al-Qaeda's commitment to acquire and deploy CBRN weapons. CBRN terrorism research would be better served by separating unsubstantiated allegations, emanating from dubious sources from facts that can further our knowledge and sense-making of terrorists' attitude toward incorporating CBRN weapons in their modus operandi. Finding methods to assess future threats and manage potential CBRN terrorism incidents requires a comprehensive, balanced and clear-headed approach.

Furthermore, projecting an unsubstantiated CBRN terrorism threat against our society's vulnerabilities may in the end become self-fulfilling. Milton Leitenberg has pointed out this aspect by referring to the documentation from Dr. Ayman al-Zawahiri's efforts to acquire a biological warfare capability in Afghanistan during the period between 1998 and 2001, where al-Zawahiri states that he became aware of the potential of biological weapons through the enemies, "repeatedly expressing concerns that they can be produced simply with easily available materials."[16] In other words, it may produce a self-perpetuating prophecy: discussions of so-called chemical shortcuts could increase the likelihood of an attack against chemical facilities/transports.

There is also an urgent need to differentiate specifics in each case to determine the exact nature of the incidents and the seriousness in intent and capability. There exists a wide array of alleged CBRN-related "terrorism incidents" referred to by government officials, media and the research community that is officially counted as projecting a general trend of an increasing interest and effort from non-state actors to acquire and deploy CBRN agents. A large part of these alleged incidents could rather be referred to as examples of disincentives and thresholds that work against these terrorists in approaching a CBRN pathway for terrorist attacks. One prominent example is provided by Ron Suskin in the unsubstantiated allegation of al-Qaeda efforts to use a Mubtakar device in the New York subway system.[17] Other examples are the "ricin cases" that are often referred to. The London Wood Green case is one example where a false negative of ricin was reported. Twenty-two castor beans, small amounts of chemicals and Arabic notes on ricin were found in the apartment.[18] Other similar cases include the Kermal camp in northern Iraq where allegations have been made of training of terrorists in poisons and chemical warfare, including the production of ricin.[19] A third example of an alleged ricin terrorism case which turned out to be false is the discovery of bottles found at Gar de Lyon train station in Paris, initially believed to contain traces of ricin.[20] A major issue is how easy it is for governmental officials and the media to publish initial uncorroborated information referring to poisons and chemicals in combination with plans to commit terrorism. To what extent are there cases where non-state actors operate out of a real commitment to acquire the necessary competence, resources, methods and logistics for producing and deploying CBRN agents for terrorism purposes? With real commitment we mean more than just initial ideas, plans or statements of an interest to pursue a CBRN pathway toward acts of terrorism. The point here is that it takes a serious effort to try dividing facts from allegations and sorting out the few cases where confirmed facts pointing toward a real commitment. To the extent where such cases exist, what can these tell us about specific incentives and disincentives for such activities? This represents a major barrier when assessing the current literature.

Another fallacy is the often nonsensical comparisons within the CBRN ladder before one considers the leap from conventional to CBRN agents. It is important to approach the issue of CBRN terrorism within a general context of international terrorism. Comparing the likelihood of chemical, radiological, biological or nuclear terrorism is of limited value without contextualizing the threat and preferred tools for terrorists, and what may drive them to include other means and approaches in their toolbox.

In this vein some argue that understanding group dynamics is more fruitful than pure focus on technical indicators – as there are too many technical pathways (BC) – every pathway has its own signature. As such there are fewer indicators to keep track of and monitor.

Nevertheless, some preceding chapters in this book provide a set of indicators and group profile aspects relevant for developing sensitivity to future potential CBRN terrorism activities. Establishing early-warning mechanisms, sensitive to activities of individuals and groups pursuing a capability for CBRN terrorism attacks, needs to include aspects of both terrorist group profiles as well as technical indicators. Given the methods of choice and capacity profiles of terrorist groups in general, these early-warning mechanisms have to be designed in order to be vigilant for waves and leaps in future technological evolution as well as societal changes.

Predicting convergence? The state of contemporary research

This book was motivated by Gary Ackerman's observation that CBRN terrorism research had reached an interpretative impasse and needed new intellectual injection to spawn new waves of inquiries. Nancy Hayden is equally scathing in her contribution, comparing the contemporary research knowledge base on terrorists' WMD motivations with a desert landscape, lacking the basic criteria for supporting decision-makers. As such, she affirms there are serious deficiencies in research methodologies which results in an unchallenged "conventional wisdom," isolation of experts within the disciplines and scarcity of structured, evidence-based reasoning.

The conventional wisdom regarding terrorists' motives for WMD continues to dominate the literature and has been challenged by new research on only three aspects: "threat convergence"; research on leaderless jihadist groups; and the evolving role of the internet in promoting global incentives and technology transfers to terrorists. Furthermore, there is an ongoing debate whether or not Islam, as a religious ideology, provides motivation or constraint.

The conventional wisdom has prevailed, according to Nancy Hayden, due to the fact that:

- there has been little re-evaluation of data to ascertain their validity in an evolving context, and therefore there have been relatively few new hypothesis being put forward to test the conventional wisdom;
- there is a small domain of SMEs and few cases for trend analysis studies. These case studies test only behavior of interest and do not provide an adequate control group for actual decisions to use or not to use WMD;
- the literature is narrowly defined to specific research disciplines and is dominated by western scholars (there is a lack of interdisciplinary approaches);
- there is a large appearance in the literature, during the last five years, of "self-referencing and repetition of ideas and authors across the majority";
- the issue of factors motivating acquisition and use of WMD by terrorists has been treated as a first order problem, when it is a wicked problem that is driven by second order, non-linear effects.

Similarly, Gary Ackerman laments that the bulk of CBRN terrorism research to date has been on an abstract level or totally lacked concrete guidance for practitioners. Researchers need to pay closer attention to requirements from law enforcement agencies, intelligence and policy-makers due to the devastating potential of contemporary terrorists. Framing operationally relevant research questions would significantly speed-up the rather leisurely pace of transition from basic research to application to policy which often occurs over a period of decades.

Insufficient data set and poor analytical toolbox

According to Gary Ackerman, to date there has never been one single case of CBRN terrorism resulting in mass fatalities. Hence, discussions regarding CBRN terrorism have a predictive character, trying to foresee the likelihood and nature of future CBRN terrorism attacks. Which aspects of the problem can we forecast in any practical sense, given that the dilemma of CBRN terrorism is a complex or wicked problem? Without such meta-analysis of the problem we might all be jumping blindly into an analytical black hole. Another impediment for predicting future likelihood of complex and dynamic phenomena such as

CBRN terrorism is the frequent use of inductive techniques by anticipating potential threats by relying on detailed data sets from past incidents and case studies: "For those areas where the past is unlikely to offer any useful guidance, we need to explore the use of new, non-frequentist and non-deterministic methods of analysis".

Nancy Hayden emphasizes the positive development of emerging public databases, facilitating a broader empirical basis for analyzing motivations (such as the GTD at the University of Maryland). Other positive developments within current research include the increasing attention paid to understanding how means and ends are interrelated and may constrain each other. However, the research community needs more refined analysis tools and approaches for utilizing these empirical data such as "social network analysis," "agent-based modeling," structured reasoning using analogues for more accurate forecasting. etc.

The CBRN terrorism research domain is heavily laden by serious shortcomings in terms of dubious data. As argued by Gary Ackerman, in order to make the most from the limited data set of CBRN terrorism incidents and relevant aspects, the research community needs to draw advantages from developments of quantitative and qualitative methods from other disciplines. Another important aspect in order to further our insights and understanding of the CBRN terrorism dilemma is the essence of combining "hard" and "soft" science. Both terrorism and technology developments are very dynamic phenomena, so in order to assess potential threats in terms of intent and capabilities, and to approach the concept of second order, non-linear research questions, we have to probe into a wider range of aspects, setting the environment and opportunities for the actors of concern.

Finally, Nancy Hayden convincingly argues that too often frameworks are posed in too narrow theoretical scope, or lacking theory altogether. Integrative frameworks facilitating cross-discipline mechanisms for exploring alternatives to WMD, not only conventional weapons but also non-violence, would provide a valuable contribution for accommodating behavior with deterring effects and at the same time support consideration of second order effects such as constituency perspective and its interaction with terrorists' choice of violent means.

More research and research collaboration between social scientists and hard scientists are needed on a number of the aforementioned critical areas. This book tried to provide a new intellectual pathway toward these ends and hopefully generated less uncertainty on how to think about threat convergence in relation to CBRN terrorism, but stimulated more questions than answers and inspiration for further research efforts in the future.

Notes

1 This was quoted in Ken Booth, "The Human Faces of Terror: Reflections in a Cracked Looking-glass," *Critical Studies on Terrorism*, 1, 1 (2008): p. 67.

2 Speech given by Chris Donnelly, "The Changing Nature of Conflict and the New Challenges of State Building," delivered at Chatham House, February 21, 2007. Available at: www.chathamhourse.org.uk/files/8267_210207donnelly.pdf.

3 Jeff Conklin, Min Basadur and GK VanPatter, "Rethinking Wicked Problems: Unpacking Paradigms, Bridging Universes," *NextDesign Leadership Institute Journal*, 10,1 (2007).
4 Jonathan Tucker (ed.), *Toxic Terror: Assessing Terrorist Use of Chemical and Biological Weapons* (Cambridge, MA: MIT Press, 2000); John Parachini (ed), *Motives, Means and Mayhem: Terrorist Acquisition and Use of Unconventional Weapons* (forthcoming); John Parachini, "Putting WMD Terrorism in Perspective," *The Washington Quarterly*, 26, 4 (2003): pp. 37–50.
5 Horacio R. Trujillo and Brian Jackson, "Organizational Learning and Terrorist Groups," in James Forest (ed.), *Teaching Terror: Strategic and Tactical Learning in the Terrorist World* (Rowman & Littlefield, 2006), pp. 52–68.
6 Ibid.
7 Ibid., p. 60.
8 Ibid., p. 61.
9 Ibid.
10 Ibid.
11 See, David E. Sanger and Thom Shanker, "U.S. Debates Deterrence for Nuclear Terrorism," *New York Times*, May 8, 2007.
12 Paul K. Davis and Brian Jenkins, *Deterrence and Influence in Counterterrorism: A Component in the War Against al Qaeda* (Santa Monica, CA: RAND, 2002).
13 Wyn Bowen, "Deterring Mass-Casualty Terrorism," *Joint Forces Quarterly*, Summer (2002): p. 27.
14 See for example: Lawrence Wein and Yifan Liu, "Analyzing a Bioterror Attack on the Food Supply: The Case of Botulinum Toxin in Milk," *Proceedings of the National Academy of Sciences*, 102, (2005); and Tara O'Toole Michael Mair, and Thomas V. Inglesby, "Shining Light on Dark Winter," *Chicago Journals – Clinical Infectious Diseases*, (2002): p. 34.
15 James Martin, "Chart: Al-Qa'ida's WMD activities," Center for Nonproliferation Studies, Weapons of Mass Destruction Terrorism Research Program (WMDTRP) at Monterey Institute of International Studies. Available at: http://cns.miis.edu/pubs/other/sjm_cht.htm.
16 Milton Leitenberg, "Assessing the Biological Weapons and Bioterrorism Threat," Strategic Studies Institute, (2005), p. 35. See also J. Littlewood and J. Simpson, "The Chemical, Biological, Radiological and Nuclear Weapons Threat," in Paul Wilkinson (ed.), *Homeland Security in the UK: Future Preparedness for Terrorist Attack Since 9/11*, (London: Routledge, 2007), pp. 74–5.
17 Ron Suskin, *The One Percent Doctrine: Deep Inside Americas Pursuit of Its Enemies since 9/11* (New York: Simon & Schuster, 2006).
18 See Walter Pincus, "London Ricin Finding Called a False Positive," *Washington Post*, April 14, 2005.
19 See statements by US Vice President Cheney in Vice Presidential Debate, October 5, 2004 and interview of the Vice President by Hugh Hewitt of the Hugh Hewitt Show, April 10, 2008. Available at: www.whitehouse.gov/news/releases/2008/04/20080410–12.html.
20 Milton Leitenberg, op. cit. (2005), p. 27.

Index